A SLIVER of LIGHT

THREE AMERICANS IMPRISONED IN IRAN

Shane Bauer

Joshua Fattal

Sarah Shourd

MARINER BOOKS
AN EAMON DOLAN BOOK
Houghton Mifflin Harcourt
Boston • New York

First Mariner Books edition 2015

www.hmhco.com

Library of Congress Cataloging-in-Publication Data
Bauer, Shane.
A sliver of light : three Americans imprisoned in Iran / Shane Bauer,
Joshua Fattal and Sarah Shourd.
pages cm
"An Eamon Dolan book."
ISBN 978-0-547-98553-4 (hardback)   ISBN 978-0-544-48397-2 (paperback)
1. Bauer, Shane — Imprisonment. 2. Fattal, Joshua — Imprisonment.
3. Shourd, Sarah — Imprisonment. 4. Political prisoners —
Iran — Biography. 5. Ivin (Prison) 6. Americans — Iran — Biography.
7. Hikers — Iraq — Kurdistan — Biography. 8. Iran — History — 1997–
— Biography. I. Fattal, Joshua. II. Shourd, Sarah. III. Title.
DS318.9.B38 2014
365'.45092313055 — dc23   2013049037

Book design by Chrissy Kurpeski
Typeset in Warnock Pro

Printed in the United States of America
DOC 10 9 8 7 6 5 4 3 2 1

For those who are not free

"Sarah, Shane, and Josh have a lesson for us all: with strength, solidarity, and a firm hold on principles, one can endure. Despite their unjust imprisonment, solitary confinement, and the looming threat of death, they stayed true to their ideals and defied the cruel illogic of global geopolitics that shackled them. *A Sliver of Light* is a richly written testament to love, survival, and the determination to make a difference in the world, whether behind bars or free."

— Amy Goodman, host of *Democracy Now!* News Hour

"In a captivating memoir, three young Americans adroitly draw us in to an engrossing tale of innocents caught not only in a foreign prison, but between two countries in a perpetual state of enmity. Writing in their three distinct voices and with no hint of bitterness — but with brutal honesty — they convey the misery and mental torture of unjust imprisonment as well as frustration at being pawns in geopolitical game."

— Hooman Majd, author of *The Ayatollah Begs to Differ* and *The Ministry of Guidance Invites You to Not Stay*

"Captures the claustrophobic atmosphere of daily life in prison . . . It's a testament to the willpower and discipline of the three captives that they maintained their values and sense of justice through their long ordeal."

— *Publishers Weekly*

"An unsugared account that demonstrates the admirable, unbreakable bond of friends, parents and countrymen." — *Kirkus Reviews*

"This engaging story portrays the horrors of imprisonment and the danger that awaits any intrepid traveler who becomes mired between the antipathy of two governments."

— *Booklist*

A SLIVER of LIGHT

The grief you cry out from draws you toward freedom.

—*Rumi*

A SLIVER of LIGHT

1. SHANE

I stir out of sleep. The air is so fresh and cool, it's almost minty. Distantly, I hear a stream purl. Sarah and Josh are lying on either side of me, unmoving. A deep predawn glow infuses everything. A bat cuts jaggedly through the air. I sit up and stretch my arms and back, which sends bursts of energy through my body. Today, we are going to hike. There are few things I love more than this.

Brown mountains jut up around us, mottled with specks of green bushes and patches of yellow grass that looks like lion's fur. The trail we started on last night snakes upward, weaving a thin little thread through the valley. We wash our faces in a nearby stream. We fill up our many little water bottles, eat some bread and cheese, and walk.

Josh is light spirited and contemplative, jumping from one rock to the next as we set off up the valley. He's so good at shaking off weariness, putting that wholesome smile back on his face. Sarah and I trail behind him, holding hands and weaving our way between the rocks. None of us speaks, except to point out the occasional curiosity, like empty goat pastures hemmed in by short walls of piled-up rocks or the occasional cement prayer niches with arrows that point the pious toward Mecca.

Hours pass as we walk. Porcupine quills, cat feces, and perfectly round spiky purple flowers appear sporadically on the slowly thinning trail. Josh is a hundred feet ahead. A cloud of yellow dust is pluming behind him, rising above the dry grass and hanging in the hazy air. Are we on a human-made trail, or did some goat slice through this endless meadow, creating this tiny track we are trudging on? The heat is growing and I am easing into that state where my body is tiring, but I just march on autopilot, pulled by something toward the top of the mountain. It must be 11 a.m. How long have we been walking? Five hours?

At some point, we stop to drink from our water bottles, which are starting to run low, and Josh mentions that we're heading east. "We could just keep going and go to Iran," he jokes. I remark that Iran must be at least a hundred miles away. We keep walking.

We reach what looks like an old, disused road, clogged with large rocks. We decide to temporarily jettison some of our things, cramming blankets and books under a bush and building a little cairn on the side of the road to remind us where the stash is. Then we plod upward, winding up the switchbacks. The ridge has to be close. The horizon — saddled between two peaks — has seemed directly in front of us for a while now. At the top, we'll turn back. We'll have to, or we'll miss Shon. He, the fourth of our group, stayed back in Sulaimaniya to rest up and is going to meet us back where we started this morning. We'll have a night around the fire before we catch a bus back up through Iraqi Kurdistan to Turkey, through the flat expanse of the Syrian Desert, and back to Sarah's and my little home, tucked into the beautiful sprawl and bustle of Damascus.

As we walk, I notice a cigarette pack on the ground. There must be people nearby. Maybe we'll find a village, have some tea, chat with the locals.

We pass an ancient-looking, broken-down stone shack on the side of the road. Sarah wants to turn back; I can feel it. Her energy is nervous, but she is trying to hide it. I'm used to this. She is strong and brave, but she's often a bit anxious when we leave cities, even when we're in the United States. She fears things like mountain lions and lone men. But she doesn't like to let the fear dictate her actions. She also doesn't like to be coddled, so I let her deal with it herself. Anyway, I want to get to the top.

"Would you rather . . . ," she starts to ask Josh and me, before trailing off momentarily. She likes to play this game when we walk and, I think, when she's uncomfortable with the silence. I love how she always starts it the same way, stating the first clause, then deciding on the second clause while the listener waits. Now she asks, "Would you rather get surrounded by five mountain lions right now, or five members of al-Qaeda?"

I think for a few seconds. "Probably mountain lions," I say. "We could probably scare them off. I think if we were grabbed by al-Qaeda, we wouldn't have much of a chance."

"Don't you think you could reason with al-Qaeda, though?" Josh says. "Speak to them in Arabic? Tell them you don't hate Muslims? Tell them you're critical of our government?"

"I don't think it would matter," I say. "But okay. I'll go for al-Qaeda. Maybe you're right. Maybe we could try to reason with al-Qaeda. There would be no reasoning with five mountain lions."

Sarah chimes in. "I would definitely choose al-Qaeda ..." She pauses. "You guys, I think we should turn back. It's getting hot and we're almost out of water."

Then, as if on cue, a tiny runnel trickles across the road. We don't have to go back just yet. The water is coming from a little spring, dribbling into a small, cement, human-made basin. I pour the water over my head by the bottleful and laugh as it runs down my skin. I can't remember the last time I felt so free. Free of time. Free of worry. Free of the heat.

Could I be more content, more happy? We take a break, our insides cooled after five hours of walking, and fall asleep in the shade. I wake to the phone ringing. It's Shon. He is on a bus and getting ready to come to meet us. How could the phone get coverage way up here? "Just go to the waterfall," I tell him. "It's right past the big campground with hundreds of people camped out. There are a bunch of tea vendors and stands selling souvenirs and stuff. From the waterfall, walk straight up the trail and up the valley. We'll be coming down soon. There is no way we can miss each other." I hang up as Sarah and Josh stir out of sleep.

2. JOSH

I could hike all day like this.

"You guys," Sarah says with hesitancy in her voice. "I think we should head back."

"Really?" Shane sounds surprised. "How could we *not* pop up to the ridge? We're so close."

I turn to Sarah, thinking of her question about al-Qaeda and the mountain lions. I think of another discussion we had, wondering if Kurdish rebels would be in these mountains of northern Iraq and how nervous she was when we were hiking last night. It seems like she's wanted to turn back for a while but kept quiet. Then I look at Shane and say, "Sarah feels strongly about this. I think we should talk it through."

I'm being sensitive to Sarah, but Shane knows me well — he knows I want to reach the top, and he asks, "Josh, what do *you* want to do?"

"Well," I say, "I think we should just go to the ridge — it's only a couple minutes away. Let's take a quick peek, then come right back down." Sarah agrees.

Just as we're setting out, Sarah stops in her tracks. She looks concerned.

"There's a soldier on the ridge. He's got a gun," she says. "He's waving us up the trail." I pause for a second and look at my friends. They seem worried but not alarmed. Maybe it's an Iraqi army outpost.

We stride silently uphill. I can feel my heart pounding against my ribs, but I want to look cool and confident. A different soldier with a green uniform and a rifle waits for us where our road meets the ridge. He's standing in front of a round, stone building that we had previously looked at and decided would be our destination. He's young and nonchalant, and he beckons us to him with a wave. He doesn't seem hostile. When we finally approach him, he asks, *"Farsi?"*

"Faransi?" Shane asks, then continues in Arabic. *"I don't speak French. Do you speak Arabic?"*

"Shane!" I whisper urgently. "He didn't ask if we speak French. He asked if we speak Farsi!"

As I speak, I notice the red, white, and green flag on the soldier's lapel. These aren't Iraqi soldiers, Kurdish rebels, al-Qaeda, or mountain lions. We're in Iran.

We follow the Iranian soldier along the other side of the ridge to a small, unmarked building. Around us, mountains unfold in all directions. There is no flag, nothing marking the building as Iranian, only a dozen soldiers in uniform milling around the building.

A portly man in a pink shirt starts barking orders. He's scruffy and he looks like he just woke up. This man in pink must be the commander. His men take our stuff. He stays with us as his soldiers dig through our bags.

He doesn't take his eyes off Sarah. He gets on his radio and communicates something incomprehensible, but still, he keeps his eyes on Sarah's body — scanning up and down. I can feel Sarah tensing up between Shane and me, and I'm getting worried.

He watches us scoot closer together. He puts down the radio and laughs at our fear. Then he walks away.

His soldiers jockey for position to examine our belongings: cameras, an iPod, wallets, a compass, a cucumber, hummus, baklava, and two books. One book is about the history of the Crusades; the other about the struggle of Iraqi Kurds for independence. They flip through our passports, talking among themselves. The only word I understand is *"Amreekaaii,"* American.

I keep asking, "Iran? Iraq?" trying to figure out where the border lies and pleading with them to let us go. Shane and Sarah do the same, talking to different men and hearing different answers. Some point to the ridge; others point to the road we walked on. Some evade the question. I scrutinize each soldier, trying to discern: How much hope should I have? How scared should I be? Most of the soldiers act reassuringly. "No worry," one of them repeats in English, "no worry."

Sarah finds a guy who speaks a little English and seems trustworthy. He points down to the ground under his feet and says, "Iran." Then he points to the road we came on and says, "Iraq." He is saying we were in Iraq when they called us over. We start making a fuss, insisting we should be allowed to leave because they called us over their border. He agrees and says in awkward English, "You are true."

It's a remote outpost and our arrival is probably the most interesting thing that has happened out here for years, but eventually the excitement dies down and the innate banality of the place returns. They feed us pasta and give us tea. Some soldiers laugh among themselves. Others try out English words on us. Shane, Sarah, and I don't have any strategy. We'll cooperate, reason with them, or argue with them, but we don't want to go any deeper into Iran than this. We want to walk down the mountain, meet up with Shon, and make our way back to Syria.

The English speaker approaches us again after having talked to the commander and says, "You. Go. Iran. You. Go. Mariwan." He says it's not up to him; these are the boss's orders. Shane and Sarah recognize the Farsi word for *boss, ra'is,* because it's an Arabic cognate.

Nothing we say matters. The wheels are already turning. Shane asks the commander if he can make a phone call. To my surprise, he actually points Shane toward cell phone reception, though he threatens to shoot him if he tries to run away.

A few minutes later Shane returns to where we've been sitting. "Shon was getting on the bus for Ahmed Awa when I called," Shane says quickly. "I told him to contact the U.S. embassy and I made sure that he knew that they'd waved us across the border. Shon got off the bus immediately. He sounded panicky."

Shon's often panicky. Nonetheless, it was his anxiety that forced us to buy Iraqi SIM cards for our cell phones so the four of us could call one another. Now, at least someone knows what's happening to us.

An SUV rattles up Iran's side of the mountain.

Sarah turns to me and Shane. "Quick, guys, what should we do?"

"Fuck," I say. "We don't have much of a choice."

"I don't think we should go in the vehicles," Shane says.

"Me neither! But what else can we do?" I say.

"We could go limp, like at a protest," Sarah suggests.

The SUV pulls up in front of us, and the soldiers start to yell at us. We don't budge.

"What should we do?" Sarah repeats hurriedly.

The soldiers close in. I say, "Let's go limp. At least, we can say that they dragged us in."

The soldiers carry Shane and me into the SUV. They don't touch Sarah, but they bark at her. She hesitates briefly, then gets in the car of her own accord.

3. SARAH

The SUV kicks up clouds of dust as the soldiers drive us down a dirt road flanked by low shrubs. For five minutes we skirt the mountain ridge; then we ease into a web of rolling hills and begin to descend.

I peer out the front window, but as we jerk over the rocky terrain, I can only catch a few glimpses through the shrubs of what lies ahead.

When we take a sharp turn onto a paved road, a startling view opens up before us. The wide, open valley is dressed in its summer skin, with swaths of pale green peppered with bald areas where dry, orange soil peeks through. In contrast to the muted colors of the earth, the sky is shockingly blue.

My brain feels divided, my thoughts bubbling over with fear and curiosity. What do I know about this country I'm being driven into? I know our governments have hated each other for decades, but part of Obama's platform was a promise to open dialogue with Iran. Last month, when it was announced that President Ahmadinejad would be reelected, huge protests erupted all over the country. Until recently, international news on Iran was dominated by footage of demonstrators, the "Green Movement," being gunned down and arrested in the streets, as well as allegations of others in custody being tortured and raped by the Revolutionary Guard.

What does any of that have to do with us? A few minutes ago, all of it was irrelevant to my life. Now, I'm being forcibly driven into a country I know little about and never intended to visit.

Still, we're nowhere near Tehran — we must be in the Kurdish part of Iran. I wonder if the Kurds are treated like second-class citizens in Iran as they are in Turkey and Syria, or if they've won a degree of autonomy as in Iraq. As my eyes scan the horizon out my window, I have an impulse to take out my notebook and write down my first impressions of the Islamic Republic of Iran. Despite my fear, it's still oddly exciting to be here.

"Sarah," the soldier in the front seat says as we drive, making eye contact with me in the rearview mirror. He motions for me to put something over my arms. Iran is the only Middle Eastern country other than Saudi Arabia and Qatar where the law requires women to dress modestly. I also know that Iranian women are known for boldly resisting this mandate. I nod at the soldier, hastily untie a long-sleeved shirt from around my waist, and put it on.

We enter a small, dusty town. When I see a few women covered in black from head to toe, I feel even more conspicuous. Suddenly, our car pulls over and stops. I glance in the rearview mirror and realize that again the driver is looking straight at me. "Sarah," he says, and points to the street. I look in the direction he's pointing but see nothing. He points again. I grab Josh's arm and lean over him to look at Shane.

"What the fuck," I say. "Is he telling me to get out?"

"I don't know," Shane says. "Just stay there. Don't move." Suddenly, my door opens and another soldier gestures for me to get out.

"No!" I shout. I grab the handle and slam the door. The soldiers in the front seat begin talking heatedly. I give Josh an exasperated look and squeeze his arm tighter. Then I reach my other arm across him to grab Shane's hand.

We watch the driver get out of the SUV and stroll across the street into a small shop. Why are they singling me out? I feel my body tensing up, ready to fight if I have to. A few minutes later, the soldier emerges from the shop. On his arm he has draped several pieces of patterned fabric. He knocks on my window and I slowly roll it down.

"I think he wants you to choose one, to cover your hair," Shane says.

Shane's right. My fear is instantly replaced by indignation as I think how unnecessary all this is. I imagine myself back in Sulaimaniya later tonight, waving the headscarf at Shon as we retell the details of our harrowing day in Iranian custody. Reluctantly, I choose a red one with green and orange flowers, turn to the driver's mirror, and carefully wrap it around my head.

4. JOSH

After our long drive down the mountain, we arrive at a police station. Inside, I am seated at a desk across from an interrogator. On my left, another man sits in a plastic chair. Beside the desk, there is a two-gallon pot of tea with a single Lipton tag hanging out. Shane and Sarah wait in the backroom, where I sat when they were each interrogated.

"What is your name?" the interrogator starts off.

"Joshua Felix," I lie, using my middle name printed on the passport to disguise my Arabic last name.

"Your religion?"

"Christian," I lie again, hesitant to admit that I'm Jewish.

"Write it down."

He tells me to sign and fingerprint the bottom of each page that I answer.

The interrogator stops thumbing my passport and says with solemnity, "We have an eyewitness." He points to the guy in the plastic chair. "He says you were in this town, Mariwan, last week. He says that he met you on the street and asked you where you were from. He says you told him you were French." The guy in the plastic chair smiles at me.

We were in Iraqi Kurdistan for the past few days and Syria before that . . . Why is this guy lying?

"I swear I'm not French. I'm American. You have my passport."

"Okay, if you are American," he says with a straight face, "then spell supercalifragilisticexpialidocious." I don't answer. I stare at him in disbelief. I feel foolish for having taken him seriously at all.

After that, he wraps up the interrogation. They take us back outside and into the SUV. They try to separate me from Shane and Sarah, but Sarah interlocks our arms and yells at them until they relent. No one tells us where we are going.

All afternoon, they've been shuttling us to different buildings. In the medley of officials I have already met in the police stations, I've found no potential allies, no one willing to listen. Everyone is lost in the bureaucratic quandary of what to do with these Americans who showed up near a border area.

In another police station, they feed us dinner, give us some blankets, and put us in a large, empty room. I lie down next to Shane and Sarah, under a mess of wool blankets. My legs feel tired from the long hike this morning. I think about my mother and father and hope they don't know what is happening to me. I go to sleep, anxious and exhausted.

5. SHANE

The next day, they wake us at sunrise and we drive. By midday, we arrive at another city and they take us into a nondescript apartment with all the blinds closed. I don't know where we are. Various people come and interrogate us with little skill, seemingly asking whatever questions come to mind. Different translators come and go. The last one says these men picked him up at the university. He says he doesn't know what's going on. He looks frightened.

After dark, a new man comes. They bring me out of the bedroom that Sarah, Josh, and I have been held in. It's apparent that they expect me to answer the same stupid questions, through this translator, for the third time today. "I won't answer anything until you tell me who you are and what is happening to us," I insist. He leans back in his chair and laces his ringed fingers over his protruding belly. "We are a nongovernmental organization," he says, grimacing.

"What. Are. You. Going. To. Do. To. Us?" I ask with rising anger in my voice. Sarah and Josh, apparently hearing my protestations, come out of the room and stand behind me. Someone else materializes and trains a camcorder on us. The man leans forward in his chair, leaning his forearms on the desk between us, and says, "In about one hour, you will know everything."

An hour later we're in a car. Beneath the night sky, the city is smearing slowly past our windows.

Who are these two men in the front seats? Where are they taking us? They aren't speaking. The pudgy man in the passenger seat is making the little movements that nervous people do when they try to pretend everything is normal: coughing occasional, fake coughs; adjusting his seating position compulsively; fiddling with the doo-dads on the dashboard with mock interest. Everyone in the car is trying to prove to one another, and maybe to ourselves, that we aren't afraid. But Sarah's hand is growing limp in mine. Something is very wrong.

"He's got a gun," Josh says, startled but calm. "He just put it on the dash." In a busy roundabout, our car swerves to avoid an oncoming vehicle. The pistol falls from the dash and scuds across the floor. My heart stops and my mouth goes dry. The pudgy man picks it up and sets it on his lap. We turn onto a road that leads out of town. The city lights fade behind us.

"Where are we going?" Sarah asks in a disarming, honey-sweet voice. *"Sssssss!"* the pudgy man hisses, turning around to face us and putting his finger to his lips. The headlights of the car trailing us light up his face, revealing his cold, bored eyes. He turns back to face the

front. The solitary lights of country houses stream by like little mete-
orites. The car falls silent again.

He picks up the gun in his right hand and cocks it three times.

Sarah's eyes widen. Her posture stiffens. She leans toward the man
in front and, with a note of desperation, says, "Ahmadinejad good!"
(thumbs-up!) "Obama bad!" (thumbs-down!) The pistol is resting in
his lap. He turns to face us again and holds his two hands out with
palms facing each other. "Iran," he says, nodding his head toward one
hand. "America," he says, lifting the other. "Problem," he says, stretching out the distance between them. He checks our faces to make sure
his message registered, then drops his arms.

Sarah turns to me and starts. What does she see? Her eyes are
penetrating. "Do you think he is going to hurt us?" she asks. I don't
know whether to respond or just stare at her. I am terrified. We walk
into our fear together, letting it surround us softly like fog. The immediate prospect of death seems so different than I had imagined it.
In my mind, I see us pulling over to the side of the road and leaving
the car quietly. My tremulous legs will convey me mechanically over
the rocky earth. I will be holding Sarah's hand and maybe Josh's too,
but I will be mostly gone already, walking flesh with no spirit. We
won't kiss passionately in our final moments before the trigger pull.
We won't scream. We won't run. We won't utter fabulous words of
defiance as we stare down the gun barrel. We will be like mice, paralyzed by fear, limp in the slack jaw of a cat. We will just stand there.
Each of us will fall, one by one, hitting the gravelly earth with a thud.

Sarah pumps Josh's and my hands. Her eyes have sudden strength
in them, forced yet somehow genuine. "We're going to be okay, you
guys. They are just trying to scare us." Yes, maybe they are just trying
to scare us. This can't be true . . .

Where's Josh? He seems so far away. His head hangs low over his
chest and he is staring blankly at the back of the seat. How did I get
him into this? I wanted the Middle East to be lovely for him. For *years*
I've been trying to get him to visit. I promised him he would love it.
I *knew* he would love it. His eyes squeeze shut as his face floods with
emotion and he tries to force it back down. Then they open again.
Then they close tight. Maybe we should choke these two in the front
seat before we become too resigned to our fate. I could get the pudgy

man and Josh could get the driver. Could we drive back to the border? How far is it? Where are we? What about the car behind us?

Our car turns and its headlights illuminate a giant, red steel door. As soon as we stop, it swings open and we pull inside a dark, empty compound surrounded by ten-foot-high cement brick walls. There are a few gray buildings inside. No one is around but the man who opened the door and another standing on the steps of the main building. Will they do it here, hidden, out of view? We all get out silently. They tell us to bow our heads and they walk us into the empty country jailhouse.

Once, almost four years ago, I was walking along a set of train tracks in Oakland with Sarah. It was summer, just before dusk, and we were drunk on our new love. We had just discussed the possibilities of our future: Could I settle down someday, have a home? Would she travel, maybe even move to the Middle East? Yes, I could! Yes, she would! We talked in that delicate way new couples do, not committing to any future plan, just dancing, each of us seeing if it was safe to take things further. As we walked, a train approached slowly from the opposite direction on the adjacent tracks. We stopped, leaned back against the frames of our bikes, and watched the graffiti-covered cars slowly roll by. The world felt perfect. The ker-klunk of the train was perfect. The breeze on my face, perfect. As I watched the groaning mass of metal, I was at once calm and excited. I thought I had given up on finding anyone who could fit into my life, who could handle the constant moving and risk taking. But at that moment, I was sure we could make it work. The train continued to roll past us and as the engine drew farther away, I heard the faint dopplering sound of a whistle. For some reason, I turned my head to the left at the sound of the third or fourth whistle. On our track, a train towered over us like a life-sized still photograph. The next thing I knew, we were off the tracks and the train was smashing by. We were in each other's arms. Our chests were heaving. Neither of us knew who saved whom and we liked it better that way.

This — now — feels like that. When the metal jail cell doors clang behind us, I fall into Sarah's arms with the same sigh of relief and the same exhilaration for being alive as I did then. The prison cell feels like life. Compared to the immediate potential of death, the uncertainty of captivity seems like a gift. Moments start to take shape again. Sarah and I kneel down, and I weep into her shoulder. Something pours out of me — a deep guilt. "I can't believe I got you into this," I say, looking into her eyes. "I could never forgive myself if anything happened to you."

She grabs my face and, almost sternly, says, "I got my own self into this, Shane. This isn't your fault. Do you understand me? This is not your fault." Part of me believes her, but most of me doesn't.

Josh sits with his back against the wall. "I want to have kids someday," he says. Sarah calls him over and the three of us embrace. And the three of us breathe.

6. SHANE

On our second morning in the jail, Pudgy returns. They are taking us back to Iraq, he says. It's all over. It must be 6 a.m.

After about two hours of driving, we pull off for breakfast at a little roadside restaurant tucked into a ravine. The pickup that has been escorting us follows. Mustachioed men in baggy Kurdish pants and cummerbunds plop casually out of the truck bed. They walk slowly to the restaurant, bandoliers hanging loosely around their shoulders, AK-47s dangling casually at their sides. Their ease puts me at ease. Our relief that we are going back to Iraq is palpable. Sarah walks aimlessly and swings her arms high and loose as she gazes up at the rock walls all around us. I meander around, plucking a little purple flower from a bush and pressing it into my book, *The Crusades Through Arab Eyes*, as a souvenir. In the restaurant, Josh snatches a tray of tea from the waiter's hands and serves it to our guards with a smile, humming Bob Dylan songs under his breath. They bow their heads, slightly but graciously, clearly tickled. We linger over our breakfasts of kebabs and eggs.

When we leave, we drive on and on, past salt flats and fields of sunflowers. Hours stack upon hours. "Iraq?" I ask Pudgy. *"Areh, areh."* He nods, *yes, yes,* and points up ahead. He does this every time we ask, as if Iraq were perpetually around the corner. Reality is slow to set in. It isn't until we have been clearly heading east for two hours that one of us finally announces that we are driving to Tehran. At one point Pudgy turns to us and holds out his palm as if balancing something delicately on it. He looks at his hand and says "Obama" as if the name were perched there. Then he blows across it — poof! — as if scattering dandelion seeds. That is his explanation for what's happening to us. For the rest of the drive, we oscillate back and forth between a heavy silence and discussing what to do if we're separated. If they pull us apart, we decide we will hunger-strike until we are reunited.

By midday we're in Tehran, where we're transferred to the custody of burly plainclothes men. They put us in a white van with tinted windows, blindfold us, and drive us for about fifteen minutes. When we stop, we are taken into a building and we sit in chairs, clutching one another's hands. Leather shoes click past. Black chadors wisp into the thin line of vision underneath my blindfold. We are made to dress in light blue fatigues and pose for mug shots, holding boards with numbers on them. The only thing keeping me from falling deep into my own fear is a strong desire to comfort Sarah. She keeps searching for reassurance that we're going to be okay. She is trembling slightly, tightly gripping Josh's and my hands.

A man tells us to go with him. We follow, hand in hand, in and out of doors and hallways, still blindfolded. Now I am trembling. Suddenly, after turning a corner, they start pulling at Josh. They've tried to separate us from him several times in the last few days, but now they are serious. I hold on to him tightly, letting my body be pumped with each of their tugs. Someone twists my other arm behind my back and I shout like an animal in pain. My brain is skipping tracks, but my body is still groping automatically. It feels sort of like when you fall off your bike, or the split second between getting punched in the face and finding yourself on the ground, when everything is black and jumbling. Amidst everything, I hear Sarah yelling, "No! Noooo!" repeatedly.

Then Josh lets go of me. "I'm going," he says. "I'm going." I feel the instant relief that comes with submission, then the loss. He floats swiftly away in the dark sea of bodies, and Sarah and I are immediately pushed upstairs. As we climb the stairs, Sarah is bent over, crying and shouting, "Josh! They took Josh!"

"It's okay, Sarah," I say stupidly. What else is there to say? "It's going to be okay."

Upstairs, they push Sarah into a cell. "Let me stay with my husband!" We aren't married, but she's hoping they'll keep us together if they think we are. "Please! Please!" she begs. "Plea —" The heavy, metallic sound of a closing door cuts off her voice.

I enter the cell across from hers submissively. The door closes. I let my back fall against the wall. Slowly, I slide down to the floor and let the weight of my head fall onto my knees. Everything is drowned in silence so thick and black that it feels like its own entity. It fills the room and squeezes up against me.

I take off my blindfold and look around. I see the carpet, a tightly knit bluish gray. I see the marble-tiled walls, gray with threads of black knotted throughout. I see the thick steel door with its little food slot and window, both sealed with their own metal doors. I see the thin plastic door that leads to the little bathroom with the little toilet, the little sink, and the snaking bidet. I am irrevocably present. It is just me and these things. Ten feet off the ground, the sun spills in through the grated windows like orange daggers. The fans in their ducts whir. The world turns with a slow groan.

Sarah's muffled sobs pulse nearby. The bottom has fallen out of whatever vitality was inside me. We lost. Sarah and Josh are gone. All they left me was Sarah's sobbing. I am grateful to them for leaving me at least that. No, I hate them for leaving me with that.

7. JOSH

My body is theirs. My sandals clap loudly on the floor as I try to catch my momentum and keep my balance. After every few steps, they spin me in circles. My mind tries desperately to remember the way back.

The door shuts behind me. The clanging metal reverberates until silence resumes. I stand at the door, distraught and disoriented. I should roar like a lion; I should cry like a baby. Now is the time to blaspheme the world, but I would be faking it. Whatever script, whatever drama I thought I was in, ends now. Whatever stage I thought I was on is now empty. The director left and so did the audience.

Slowly, as if I were sick, I dodder to the corner of my cell and take a seat on the carpet. There is nothing in my eight-by-twelve-foot cell: no mattress, no chair, nothing — just a room, empty except for three wool blankets, with a bathroom attached. My prison uniform blends in with the blue marble wall behind me, and the tight blue carpet below. Shane and Sarah are probably sulking in the corners of their cells too. We agreed we'd hunger-strike if we were split up. Now I don't feel defiant. I just feel lost.

Sarah's glasses are in my breast pocket. She gave them to me to hold for her when they made us wear blindfolds. She didn't have pockets in her prison uniform. I empty my other pockets: lip balm from the hike and a wafer wrapper — the remnant of my measly lunch.

A creeping sense of aloneness takes root. I don't know what I'll do in here for the rest of the day. Beyond my bewilderment, I fear the encroaching emptiness. I sense the hovering *blankness* — a zone of mindlessness that looms over my psyche and lives in the silence of my cell.

8. SHANE

The day after our arrival, men remove me from my cell. They seat me in a chair-desk combo — the kind used in high schools — facing the corner of a room. I'm blindfolded. I don't know where I am, except that it is in or near Tehran. I'm convinced we're in an unassuming building in some alley — a kind of secret prison. We must be in an outlying neighborhood. Why else would I hear birds chirping?

I can hear a group of men whispering behind me. Chairs are shuffling. The sweet pungency of dried leaves hangs in the air. My oversized blue prison shirt exposes my chest. The air is cool, but not sharp.

Last night, Sarah and I discovered we could whisper to each other in short bursts through the grates on the bottom of our doors when no one is nearby. Neither of us has heard from Josh since yesterday. We haven't eaten since we arrived.

Things are moving forward. We are going to clear this up. They will ask me questions and they'll understand they made a huge mistake.

*They don't want an international scandal. They are figuring out how
to ease out gracefully.*

"How are you?" a voice melts in softly. "The interrogator is going
to ask you some questions. I will write them down in English and you
will write your answers." I am trying to act relaxed, nonchalant, and
confident of my innocence. The interrogator makes brief, bullish ut-
terances. The translator hands me a piece of paper.

Q: What was your mission in coming to Iran? Who sent you here?

"I didn't have any mission," I say. "I never meant to come to Iran.
Your guards waved us over the border —"

"Please just write your answer."

*It was either Iraqi Kurdistan or Lebanon. Sarah had a week off
work at the American Language Institute in Damascus, where we'd
been living for a year. Shon and Josh were visiting. We wanted to take
a trip. Sarah and I loved Beirut, but it was just a few hours away. We
could do that anytime. Why not go somewhere farther away, somewhere none of us had been? Sarah had wanted to go to Iraq for a while.
I'd been to Baghdad and Fallujah as a journalist, but never to the
"other Iraq." Kurdistan is a different place, almost a different country.
I've read write-ups on Kurdistan as a tourist destination in several
publications, including the* New York Times *and* Vanity Fair. *About a
million tourists go there every year.*

*It is one of the only pro-American parts of the Middle East, more
so, I would say, than even Israel. Iraqi Kurdistan, one security contractor told me while I was in Baghdad last February, is where people
like him go on vacation. That makes sense — the United States virtu-
ally handed Iraqi Kurds their autonomy. It was the least the Americans could do after what happened in 1991. That year, George H. W.
Bush made a radio broadcast encouraging Iraq's Kurds and Shiite Ar-
abs to rise up against Saddam. Apparently believing they would be
supported by the U.S. military, both peoples revolted. Saddam killed
thousands of innocents and rebels. During the bulk of the massacre,
the United States stood by. Eventually, the United States enforced a
no-fly zone in the north and south of the country, ostensibly to allow
the return of Shia and Kurdish refugees who fled the Iraqi military's
killing spree. The Iraqi troops withdrew from the north and a de facto*

Kurdish regional government was established, creating an autonomous region.

We visited Kurdistan because it is nothing like Iraq proper. At the border between Turkey and the Kurdish region, officials happily stamped our passports with tourist visas. To get a visa to Baghdad five months earlier, I had to transfer money from Damascus to my fixer so he could bribe officials. In Kurdistan, we visited a castle and ate pizza on the streets. We watched the city of Sulaimaniya erupt in festivities after Kurdistan's parliamentary elections. I witnessed elections in Baghdad months earlier and the streets were dead. People were afraid of bombings and shootings. Driving was forbidden. In Kurdistan, there were fireworks and music and dancing. It was beautiful.

It seemed like everywhere we went during our first two days in Kurdistan, there were pictures of fantastic mountains on display, so when we arrived in Sulaimaniya on our second night, I asked our taxi driver if there were any places to hike. The place to go, he said enthusiastically, was Ahmed Awa. When we asked our hotel manager the same question, he gave the same answer, pointing to a large poster vaguely reminiscent of a Swiss hamlet with snowcapped mountains and poorly photoshopped waterfalls. "Ahmed Awa looks just like that," he said. Next to the poster, taped to the wall, was a picture of a waving European-looking couple standing in front of a waterfall at the actual Ahmed Awa.

Sarah, Josh, and I wanted to go and camp overnight, but Shon didn't. He thought it was a bad idea and, anyway, he wasn't feeling well and wanted to spend more time in the city. He'd meet us there the next day, he told us.

We arrived at the waterfall at night. It was underwhelming — more a small stretch of rapids than a cascade — but in the Middle East, water is always an attraction. Hundreds of people were camping out with their families. Men played backgammon and poured tea out of thermoses while little children chased one another in and out of tents, squealing with laughter. The air smelled of the apple-flavored tobacco smoke of water pipes. Vendors sold tea and kebabs everywhere.

We asked a tea vendor if there was a trail we could hike on. He pointed us to a wide, well-used path hugging the stream. We walked by moonlight, hearing the music and laughter from below as we *stepped carefully over protruding tree roots. The beginning of the trail was lined with stands, closed because of the late hour, displaying cowboy hats, flashlights, and those packets of crappy dry cookies for sale. We walked for a while, then found a soft spot to stop for the night.*

In the morning, we hiked. The tea vendor probably didn't expect we'd walk for hours.

A: We never meant to come to Iran. We had no mission in Iran. We had no mission at all. We were tourists in Iraqi Kurdistan. Our taxi driver and hotel manager recommended we go there. The driver's number is in my phone. You can find the hotel. Call them. They'll tell you.

I hand the paper back over my left shoulder. He reads it aloud in Farsi. The tongues of several mouths tut loudly in disapproval. He hands me another sheet of paper.

Q: What American officials did you meet with in Iraq?

There were those teenage soldiers on the Iraq-Turkey border I talked to with Josh and Shon, but I wouldn't call them officials. They were sitting, looking bored on their Stryker.

"Where are you from?"

"Mississippi. You?"

"California."

"California?! What are you doing here?"

"We're on vacation."

"Vacation? You came here for a vacation?"

A: None.

More tutting.

Q: Iraqi officials have said that the police warned you not to come near Iran. And an Iraqi newspaper said that you were spies.

The spy part doesn't surprise me. I've been detained for short periods a number of times in the Middle East on suspicion of espionage. It's the default allegation against any foreigner doing something unusual. In Iraq, each political faction has a newspaper, and newspapers in the Middle East aren't exactly known for checking their facts. But what is this about the police? Did he make that up, or did the local police department quickly drum up a story to cover its ass?

A: The word *Iran* was never mentioned to us during our three days in Iraq. Police never warned us about anything.

Q: You were carrying with you a compass, a map, a GPS, and a professional camera. Why?

A: I am a photographer and I carry my camera everywhere I go. I take pictures all the time. I was taking pictures of our hiking trip. Josh had a compass on him from his travels before. I didn't have a GPS. I think you are talking about my iPod. That is a music player. We didn't have a map either.

Q: How do you know Josh?

We met six years ago. We were both twenty-one. It was summer, and I had just come back from Yemen, where I had been studying Arabic, and found myself in the backyard of a West Oakland house with tall unkempt grass and fruit trees bursting with figs and lemons. I sat around a picnic table with other people, and we ate pasta with fresh basil and drank cheap wine. Could I park my RV in the driveway of their collective house and live there? Sure, they said.

A few days later, I drove the lumbering thing in. Inside the house, I found some people holding carrots and chatting over a bucket of peanut butter. I grabbed a carrot and dunked it in the bucket. One of those people had a big laugh and a gentle demeanor and sang Bob Dylan under his breath often. He was so obsessed with Gandhi that he would sometimes drink olive oil straight because he read that Gandhi used to do it. In that kitchen the two of us would critique the United Nations and discuss the Middle East and how to manufacture peanut butter. We would go on hikes and camp on the beach. We would trade books and dance crazily.

A year later, he would move away, off to a sustainable living community to live and work at an environmental education center in Oregon and I'd go visit him in what seemed like a paradise after crack dealers started preying on our house in Oakland. We'd pick vegetables from his garden, gather duck eggs, and talk about our lovers and our projects. As the years passed, we grew and changed. He went deep into the land and his little town, and I spread out into the world, making a modest living by selling stories to newspapers and magazines. I'd try for years to get him to come to the Arab world. Eventually he would and he would like it; then he would end up here, in Iran, in prison.

A: Josh and I lived together when we were in college.

Q: How do you know Sarah?

I think it started with a cup of Turkish coffee. I preferred to add a little sugar before the water boiled up—it brought out the coffee's sweet notes and I wanted those notes to be delightful, but subtle. I served her the coffee on a little Bosnian tray, in my kitchen, pouring it from a tiny copper pot into a tiny cup. As we drank it that first time we met, we talked about the Middle East and her work with the Zapatista rebels in Mexico. We laughed a lot.

Or maybe it really started when we walked around a Bay Area suburb, along the border region where the city's growth was checked by swampland. We walked out into the reeds and I pushed her back. She grabbed hold of me and pulled me down with her and we kissed. It wasn't the first time we kissed, but it was the kind of kiss that makes you know that something is starting that will change you.

A: I met her in my early twenties. We went to antiwar protests together.

Q: What is your e-mail address and password?

A: I'm not giving you my e-mail password.

They put a stack of nine empty, numbered pages in front of me. "Tonight, when you are in your cell, we want you to write your complete biography."

A guard grabs my arm gently to escort me out. The translator stops us. "I hope this will be cleared up very soon," he says. "If it is true that you have protested against your government, then we want you to be free. You are a human being like me. I am going home to have dinner with my family and you deserve to do the same. I hope I don't see you again." He chuckles softly and puts his hand on my shoulder. The guard leads me away.

9. SARAH

"Sarah, eat this cookie."

"Not until I see Josh and Shane."

I'm sitting blindfolded in a school chair. A cookie sits on the desk in front of me. The interrogator is saying something to me—what is it? I have to force myself to focus.

"Do you think we care if you eat, Sarah?"

They do care. I know that much. I've been on hunger strike since they split us up two days ago. At first it was difficult, not eating, but I'm learning how to conserve my energy. When I stand up, my heart beats furiously in my chest, so I lie on my back on the floor most of the day, sleeping as much as I can. Terrible thoughts and images occupy my mind — my mom balled up on the floor screaming when she learns I've been captured, masked prison guards coming into my cell to rape me — but I've found ways to distract myself, like slowly going over multiplication tables in my head. I do this for hours at a time, starting with the twos and going up into the teens until I have to stop and start over again. When it becomes impossible to continue, I sing songs to myself in a whisper.

"Sarah, why did you come to the Middle East to live in Damascus?" the interrogator asks. "Don't you miss your family? Your country?"

"Yes, of course I do. But it's only for a couple of years. I can't believe you're asking me this — do you realize how scared and worried my family must be? It's horrible what you're doing. Why can't I make a phone call and tell them I'm alive?"

It kills me that I can't even comfort my mother. When I was six, she put everything we had — my toys, books, clothes, blankets, and the cat — into a small rental car and drove us from Chicago to Los Angeles. My dad loved me and had always been sweet to me, but their marriage was volatile and our home was often an unhealthy place. When bad arguments between my mom and dad began to lead to violence and injuries to my mom, she knew we had to leave. We had to start fresh.

My world exploded on that trip. Snaking through narrow canyons and valleys of wildflowers set me on fire. I remember arriving at the Grand Canyon, jumping out of the car, and running straight up to the edge. As the beauty engulfed me, I could feel my mom's fear and hesitation tugging at me from behind like an invisible string. She never knew whether to discourage my wildness or not. She wanted to let me run free, to trust me. More than that, she wanted me to trust myself.

I snap back to the present. The interrogators have returned to the room. There are four or five of them, and they come and go so often, I'm only vaguely aware of their presence or absence. One of them — he seems like the boss — is pacing and talking angrily in Farsi. He seems to circle in close, then recede, as he continues ranting. They tell me if I eat their cookie, I can see Shane and Josh.

"Let me see them first — then I'll eat."

"Sarah, you say you are a teacher. Have you ever been to the Pentagon?"

"No, I've never even been to Washington, DC."

"Please Sarah, tell the truth. How can you be a teacher, an educated person, and never go to the Pentagon? Describe to us just the lobby."

"I've never been to the Pentagon. Teachers don't *go* to the Pentagon!" I want to ask him if he's ever left Iran, or if he even has a notion of what Iran is like outside this paranoid cabal. At times, their questions are so absurd, I almost have to stop myself from laughing, partially because I'm weak from not eating and partially because I can't really convince myself this nightmare is real.

"Now, tell us who sent you to Iran. Who do you work for? Who pays you to go on these missions?"

"I never intended to come anywhere near Iran. You know this! We don't even speak Farsi. Would a female spy be stupid enough to waltz into Iran without a headscarf?"

"Sarah, you crossed our border illegally. We need to know why."

"The border isn't even marked. We didn't know we were near it. Do you even know exactly where it is?"

"Of course we do, Sarah."

"Then why don't you show me a map? If the border is the road, then your soldiers called us off that road, into Iran. We didn't walk in by choice."

Their story is ridiculous. Spies would have to be suicidal to enter Iran the way we did, without supplies, visas, or even basic knowledge of Farsi. Any suspicion they have will disintegrate once they verify what we're telling them, right?

The guards always take Shane and me out of our cells to be interrogated at the same time. We've spent hours discussing our interrogation strategies. A few days after we were brought here, I discovered a plastic tube sticking out of the ceiling in my bathroom. When I stood up on my sink and spoke into it, the sound traveled to Shane's cell, emerging from a similar tube that opened into his bathroom. I have to stand on the tips of my toes and cock my head at an almost impossible angle to reach it, but it works far better than the grate at the bottom of our cell doors we used at first. Also, it isn't as close to the hallway where the guards walk by—so hopefully we won't be caught.

They ask me to draw a map of our apartment in Damascus. "Sure," I say. "Good idea. Why don't you go there? You will find out that everything I've been saying is true. You will find a stack of ungraded papers on my desk, bookshelves full of books that criticize the U.S. government's policies, criticize Israel. I'm sure the Syrian government would have no problem with you searching it."

I hand over the dozen or so pages I've carefully written. Like switching channels, I tune out their conversation and listen to the birds outside the window. I grew up strong, like my mother wanted me to. It was hard to get by on a nurse's salary and we struggled—but my mom made a good life for us in warm, beautiful California. I went to good schools, had lots of friends—but like any parent, she couldn't protect me from the world forever. When I was sixteen, I was raped while on a date with a slightly older guy. It took me years to pull myself out of the self-loathing and depression that violation caused. I learned to love myself again, but the trauma of being a rape survivor follows me everywhere.

Now, in prison, I have a hard time falling asleep because I'm afraid someone will come into my cell and catch me off-guard before I can defend myself. Even on the hike before we were captured, I was preoccupied by the thought of coming across soldiers or lone men on the trail. My fear seemed to contradict my objective knowledge of how safe and popular Iraqi Kurdistan is for tourists. I told myself I was overreacting and I still think I was, but maybe my fear was telling me something, like that we were simply too far from the other people camping by the waterfall, that we hadn't seen a map detailed enough to tell us exactly where we were. Maybe I should have listened.

The translator stops speaking and there is silence. Then, I hear the slow, deliberate rasp of paper being torn. The interrogator throws the papers in a pile at my feet. I feel myself emotionally detaching from the situation; his angry voice sounds farther and farther away, in the background. In my mind I can see myself from a vantage point, high up on the ceiling, sitting blindfolded in this small wooden chair with four or five angry men circling me like vultures. Maybe every moment of my life, every hardship and challenge, has been in preparation for this. If my mother could talk to me right now, I know what she would say. She would tell me that I'm stronger than these assholes, that they can't hurt me. All I need to do is believe in myself, in my ability to get through this, and I will.

"He says this is not useful to us, Sarah," the translator says, almost apologetically. "You will have to write it again."

10. JOSH

"Deplorable means bad." I answer my interrogator, "When I say Israel's treatment of Palestinians is deplorable, I am saying Israel treats the Palestinians badly." I'm acting civilly, but I want to scream at them, "I'm not the enemy you think I am!" Why are they asking me about the Palestinians? We both oppose American support of Israel. But these guys don't care. To them, I'm an American, and America threatens to attack Iran. America slaps sanctions on them, supports Middle East dictators, bombs their neighbors, and arms the Israelis. To them, I am my government.

In one sense, they are right: I am their enemy. A couple months ago, there were huge protests throughout Iran. At the time, I was in Sweden visiting my brother. Stockholm's sizable expatriate Iranian community protested in solidarity with the uprising in their home country. My brother, Alex, and I documented the anti-Iran rally in Sweden. I've been praying in my cell that Alex doesn't e-mail me about it. If he doesn't know I'm detained, he may send me his next version of the video.

"Why did you come to the Middle East?" the interrogator asks.

"I came to visit my friends."

I give that answer, doubting it will suffice for my interrogators. It didn't even satisfy my parents. When I told my father that I planned to visit Syria, he asked sarcastically, "Are you going to meet Hamas in Syria? Hamas's leadership is in Damascus."

He was prodding me. My father interprets my pro-Palestine stance as a personal rejection of him. We can argue logically about the Israeli occupation, and he'll even agree with me at times. In the end, it always comes back to his fear that I don't love him and his side of my family enough. We've ruined too many family outings debating Israel/Palestine, and I wanted him to know my trip to Syria wasn't meant to hurt him.

"Dad, I'm not going to Gaza — I'm just visiting Syria. You know I don't care for Hamas, just like I don't care for the Israeli government."

"Okay, Josh, just tell me: are you going to Syria to spite your father?"

"Dad! A part of me is going there because of you, not in spite of you. It is through you that I've been interested in the Arab world."

He was born in Basra, Iraq, in 1948 and was airlifted to the new State of Israel in 1951. These airlifts — known as Operation Ezra and Nehemiah — relocated my father as a toddler, along with 125,000 others, 95 percent of Iraq's ancient Jewish community.

His childhood memories took place in Israel, but they smell like home-cooked Iraqi food, and they sound like classical Arabic music. By age sixty, he now has lived in America most of his life, but he wants to be buried in Israel when he dies. One of the reasons I liked the idea of visiting Iraq was to see the country where my father was born.

I heard him take a couple deep breaths over the phone before responding, "Okay, okay. You know . . . Damascus is supposed to be very beautiful. You're going to have a great time. But please keep in touch. You must tell me about it — the music, the parties, everything. I would probably love Syria, but I'm not allowed there. You know, I grew up with a lot of Arab Jews from Syria, Halabis. Be sure to visit Halab and send me some photos, would you?"

"I'd love to, Dad."

The interrogators want more from me. They press, "So, you visit your friends even if they are in Syria? Don't you have friends in America?"

My mother similarly questioned traveling to Syria.

"Mom," I told her, "Shane and Sarah have been there for a year. It's safe."

"But you're Jewish!" my mother chided me over the telephone.

"Why do you think that Syrians are anti-Semitic?"

"They are not Jewish! Sarah is not Jewish, is she?"

"No."

"Exactly!"

My mother is probably freaking out right now. She expected to hear from me every few days while I was in the Middle East. I didn't even tell her I was traveling to Iraq. I figured I would mention the trip after I safely returned to Syria. I wish this weren't happening to her.

To my translator, I explain why I was traveling abroad. I want them to understand me. If they can see who I really am, maybe they'll release me. At the least, they'll have sympathy for me even if they don't have the power to free me. Hoping to ingratiate myself with them, I also emphasize my critical views of the United States.

"I was teaching for an undergraduate study-abroad program for American students during the spring semester. I traveled and taught about public health in Switzerland, India, China, and South Africa. I encouraged my students to be critical of the medical establishment and to analyze the links between capitalism and illness. The program focused on the social determinants of health, not just germs. We discussed lifestyle choices, diet, alternative medicine, and environmental quality as ways to promote public health. When the job ended in May, I wanted to visit friends abroad before returning to America."

I don't want to talk about my interest in the Arab world because I want to avoid talking about my father and international politics. I'd rather not mention the word *Israel,* which I've visited a handful of times, even if I'm criticizing it. I avoid my heritage. It feels like my original sin.

My interrogators leave me alone for a moment as they converse. I cannot attach any meanings to their Farsi words, but I remain attentive, hoping for a clue to what they're saying. One interrogator repeats my name slowly. "Jo-shua Fat-tal, Yo-shua Fat-tal." Damn my name! Yoshua was an Israelite spy in the Bible. Fattal is Arabic. One interrogator concludes, "Josh Fattal, *Yahudi Arabi.*"

Now they know. They ask me, "What's your religion?" I knew the Christian façade wouldn't last. But am I really Jewish? I don't believe in the Torah, in Adam or Eve, or Noah or Moses. I don't keep kosher, and I don't obey the Sabbath. Judaism is a religion; it is a choice. It doesn't run in my blood or anything tribal like that. Sarah and Shane don't consider themselves Christian. Why should I consider myself Jewish?

But I can't help it. I feel Jewish. Maybe I shouldn't hide it. It's the truth. But I feel like I'm confessing when I answer their question: "I am Jewish."

11. SHANE

My hunger is my strength. Three times a day, the slot on my door opens and a tray of food is handed in. I take it and hand back the full tray of food from the last meal. At first, I try to block the entry of meals with my hand, but the guards always insist, so I eventually submit. But they can't make me eat it. Today is the third day, and I've decided to start hiding the food from myself in my bathroom. The smell makes me hungry, not just in my stomach but in my bones. But I can tolerate that hunger. I can tolerate the dizziness. I can handle the momentary blackouts when I stand up too quickly. Whenever I hear, or think I hear, the guards' footsteps, I rush into my bathroom, grab the food, set it by the door, lie on the floor, and suck in my stomach to accentuate my ribs.

The hunger gives me something to focus on, a purpose. Everything I do is a strategy, to beat the hunger and to beat them. I focus my attention on not walking, on conserving every bit of energy I can. I keep my stomach full of water at all times. I refuse to take even a nibble of the food, because I think it will bring my stomach back to life.

I spend my time on my back, surrounded by the objects of my entertainment. I turn a two-liter bottle from side to side, watching the water flow from one end to the other, making little waves. I open and close the cap of my toothbrush. Click, click-click, click-click-click-click. I fold up bits of paper cup and shoot them at the window with a rubber band.

I eagerly await mealtime so I can have the satisfaction of noncompliance, the little feeling of victory to carry me through the following hours. Sometimes after refusing, I can't help pacing laps around the cell. The refusal brings me to life, makes me feel the strength in my slowly weakening muscles. I know deep inside I can't beat them, but I need the fight itself to keep me going.

12. SHANE

I'm back in the interrogation room. As I sit, waiting for the interrogator, I hang my head dramatically to suggest I am fading from hunger.

"I suggest you cooperate with them," the translator says to me.
"They are being nice now, but they can use other methods to make
you answer their questions. They say you should not complain — your
situation is easy. Do you know that there are five Iranian diplomats
being held by the Americans in Iraq? Diplomats! They have been
holding them for more than two years."

The interrogator enters and immediately asks me, again, for my
e-mail address and password. I immediately give him the password
for my two accounts. I want this to end as soon as possible. I have
nothing to hide anyway, so why not be an open book?

Q: We have proof that you have connections with U.S. intelligence
agencies. You have been in contact with groups like the Center for
Defense Information and the Council on Foreign Relations. These
are very secretive organizations. What is your relationship to them?

A: These groups are not intelligence agencies. They are think
tanks. I interviewed them for articles that I wrote. I don't have any
special relationship to them. I simply went to their websites and
clicked on the "Contact Us" tab.

"These articles were only a cover-up," the translator scolds. "This
is a very common strategy with the CIA."

Q: If you are such a peace-loving journalist, would you be willing
to write an article that we assign you?

A: I could not do anything for you while you are keeping me in
prison.

I want this to be over. I want to go home.

"We know that you had connections to Palestinians in Damascus,"
the interrogator says. "We know you met with Hamas. We know you
have made phone calls to officials in the Israeli government."

"Yeah, I interviewed them for articles during Israel's bombardment of Gaza," I say. I am getting exasperated. "Go online. You can
find them."

Questions follow questions. He tells me to list every Palestinian I have ever met. I name a few officials of Hamas, and leave it at that. I want to stay focused, but this has been going on for so long, it's getting hard for me to stay sharp enough to think through the possible implications of every answer before writing it down. I don't know how to engage with them anymore, especially when every aspect of my life in the Middle East is being formed into evidence of espionage.

Q: We know you lived in the Palestinian neighborhood Yarmouk. Why did you live there?

I moved to Yarmouk because the falafel there was like nowhere else in Damascus. I loved how the streets felt lived in — how they filled every night with scraps of fruit rinds and newspapers and plastic bags yet were clean by the morning. I loved how Ahmed cooked. I loved how Omar talked about poetry and books. I loved how Yamen could listen like no one else and had wisdom beyond his years. I loved the nights at Mazen's — the dancing, the discussions of war and justice, the laughs, the sound of ice cubes in glasses, and the sweet, anise sting of araq. Mazen did five years as a political prisoner, but he didn't talk about it unless asked.

I loved that hour in Yarmouk, which I've found nowhere else, at about 4 a.m., when pious old men shuffled to the mosque to begin the day and clutches of young people walked home with a swagger to end their night. I loved that everyone lived so close together — more than 112,000 people in 0.8 square mile! — that the sounds of music blended with the sounds of gossiping women and kids playing soccer in the alley and pigeons cooing in our windows. I loved to sit and type my articles about Iraq or Palestine or Syria to the accompaniment of those sounds.

I loved the daily conversations. Is Obama better than or the same as Bush for the Middle East? Is Faulkner better in Arabic translation than in English? Should secular Palestinians back Hamas and Islamic jihad — groups they despise — in the face of Israeli bombardment?

If only you could have seen our last night in Yarmouk, before we left for Iraq, then you might understand why I lived there. If you could have seen Emily and Basel's wedding, the way we all paraded down the streets — Palestinians, Syrians, Iraqis, Americans, Italians, Brits, Nigerians, Poles — from the ceremony to our apartment. We poured onto Yarmouk Street and our Palestinian friends jubilantly chanted

"Down! Down with Hamas!" as bearded onlookers watched, discomfited, from the sidewalk. I know if I told you this, you would say these particular Palestinians were obviously agents of the West, influenced by their Western friends, but the chants were their own and they despised the U.S. and Israeli governments at least as much as anyone else in the camp.

If only you could see into our apartment that night — the way Magda and Abu Hashish's hips cut sideways into the music and the sweaty air as their hands drifted above their heads. The way the smoke hung on people's skin. The bowls of hummus and lentil soup. The way that, after everyone left or drifted off to sleep, Shon and Yamen and I couldn't let the night end. "You don't need to go," I told Yamen. "I know. I know," he said, patting my knee. I'm not going anywhere. Let's just enjoy this moment.

The mournful notes of Umm Kulthum drifted in from a courtyard across the alleyway. We listened and smoked in silence. She was telling us that the night was over and that we should mourn it as endings deserve to be mourned.

I moved to Yarmouk because it was so wonderfully gray. The buildings were gray. The alleys gray. The blacks and whites — the tones that you hold so dear — dissolved in Yarmouk. I loved Yarmouk because it was the antithesis of people like you.

A: We lived in Yarmouk because it was cheap and because Sarah's Arabic tutor lived there.

Without skipping a beat, he hands me a piece of paper with a picture of a crowded street. There is an arrow pointing to one random person's head in the crowd.

"Where is this and who is that person?"

"This is one of Sarah's pictures from Sulaimaniya. I have no idea who that person is."

It feels like we are enacting some kind of cheap spy movie. He has looked over the pictures on Sarah's and my cameras and is acting like they hold the secret to an extensive plot.

"What is happening in this picture?"

"That is Sarah and her friend walking to a mosque in Damascus."

He hands me another picture.

"That is Sarah with her students."

"That is Sarah having a picnic with her friends."

"That is me, Sarah, Josh, and Shon at the Iraq-Turkey border with the sun setting behind us."

He hands me another picture of a bunch of young-looking shoulders with guns hanging off them. *Oh God. That's Tel Aviv. Shit. They know we've been to Tel Aviv.* The image is from Sarah's camera. She had erased the pictures from our Israel trip, but they must have recovered them using special software. Suddenly, I feel like I've been caught hiding a real crime. I've been hoping to keep it secret that any of us have ever been to Israel. Even in normal circumstances in the Middle East, like living quietly in Syria, a visit to Israel is grounds for permanent expulsion from the country. I can only imagine what it could mean in our current context.

"Those are Israeli soldiers at a bus station in Tel Aviv. We went there to visit our American friend Tristan Anderson, who the Israeli military shot in the head with a high-velocity tear gas canister while he was protesting with Palestinians in the West Bank. He was in the hospital."

"Write it down," he says.

I write for what seems like an hour. Then he hands me another picture that is slightly blurred. It is a picture of a missile. *We're fucked.* Sarah took this picture out a bus window in Israel, somewhere between Haifa and Tel Aviv. "What?" I remember her saying as she snapped the monument. "They have a *missile* on the side of the road?" This image is the perfect piece of spy evidence for our captors: the Americans were in Israel, doing their military work; then they came here to spy on us.

It's hard for me to remember that I don't actually believe that talking to Palestinians or visiting Israel or Palestine is a crime. I am starting to feel actually guilty.

13. SARAH

One evening four days after we arrived at this prison, the guards lead me to the interrogation rooms. They tell me to take off my blindfold. I see Shane seated in a tiny room dominated by a large mirror. On a small table next to him sit two plates of beans and rice.

"Where's Josh?" I ask.

"I don't know," he says. "They just told me to sit here and wait."

"He's already eaten," says a voice behind the mirror. It's our translator.

"That's a lie. Where is he?" Shane asks.

"We're not eating until we see him," I say.

"It's too late for us to bring him tonight — we need to go home to our families, Sarah. If you eat now, you will see Josh after the weekend. If you don't eat, you won't see him at all."

Shane looks exhausted, slouched over in his ugly, light blue prison uniform. Neither of us has eaten in four days, but I know I'm over the initial hump and can keep going. Faced with this cruel, unnatural choice, I don't know what to do.

"We'll see him tomorrow?" Shane asks.

"Yes, if you eat," the voice answers. Shane and I pause, looking into each other's eyes for the first time as prisoners. Reluctantly, we agree, but we tell the translator we'll stop eating again if he reneges on his promise.

I push down the nagging fear that we've made the wrong decision and fall limp into Shane's arms. My lips find his lips, his hands grab my hands, and our foreheads come together like two palms in prayer. That's what I think of — prayer. His touch feels sacred.

"I love you," I say.

I glance at myself in the large mirror next to us. *God,* I think briefly, *I look as bad as Shane does.* My black hijab makes me look severe. The dark circles under my eyes seem to cut into the dry, splotchy skin around my cheeks. How could I have already changed so much? Suddenly, I hear a brief scraping sound like the hard soles of men's dress shoes against concrete. The sound is coming from behind the mirror.

"Why are you watching us?" I ask. "Can't you give us some privacy?"

"Sarah, you have three more minutes. Please hurry."

Two days later, back in my cell, I glance toward the window and note the gray-blue sky. It's almost dinnertime. I hear Shane cough in the hallway — the signal we send to each other when one of us is taken out of our cell. My head jerks up and I look around, wondering how long I've been sitting in a semi-catatonic state, staring at the wall with unfocused eyes.

They must be taking us to see Josh. I jump up and start pacing the room, feeling the initial excitement of activity succeeded by doubt and unbearable tension. I order myself to take ten deep breaths. One, two, three . . . Then my cell door opens, and a female guard hands me my blindfold and motions for me to get dressed. They lead me to the same small, mirrored room. When I take off my blindfold, I see Josh and Shane sitting next to a table stacked with food.

"Josh," I say, throwing my arms around him and kissing his cheek, "I'm so glad to see you!"

"Don't talk too much," a voice says from behind the glass. "Only eat."

Josh smiles at me and we all begin to eat eagerly, chomping down on juicy beef kebabs that we wrap in flatbread and dip into some kind of creamy yogurt dip, washing everything down with orange juice. As we eat, we share snippets of information under our breath. Josh tells us that he's just now breaking his fast.

"They lied to us," Shane says. "They told us you'd eaten."

"At least we know the hunger strike worked," I say, making a mental note never again to take the interrogators at their word. "I think they might let us out soon," I continue, "and they clearly don't want us to look famished when they do."

"Josh, are you okay with asking them to fly us to Beirut?" Shane asks. "It will be a good place to feel things out and see if it's safe to go back to Damascus."

"Josh, you'll love Beirut," I say.

Josh's eyes are red, like he might have been crying before they brought him here, and he looks tired. Looking at him makes me feel uncomfortable — almost responsible for his pain. I can't imagine how he's been surviving this last week alone.

"Sarah, isn't it your birthday in two days?" Josh asks.

"Yeah, thirty-one. I really hope we're not here," I say.

He places something in my hand under the table. I use my fingers to identify the two objects, my glasses and a tube of lip balm. "Happy birthday," he says, and smiles.

"I can't believe you still have these. Josh, are you really okay? We've been so worried about you."

Before he can answer, the guards are back, snapping their fingers and ordering us to get up. Shane and Josh are up ahead of me. Shane angrily turns his face toward the interrogator and says, "When will we see each other again? If we don't meet in two days, we're going to start hunger-striking again."

There's no reply.

14. JOSH

Two days later I stop eating because they don't let us see one another. It's now been six days since then and I still haven't eaten. I assume Shane and Sarah are fasting too, wherever they are.

I think I could stand up if I really wanted to. But why waste the energy? I try to conserve every possible calorie. I used to feign weakness in front of the guards to gain sympathy or to make them fearful I'd become sick. Now I still tell myself that I am pretending to be tired, but, somewhere along the way, pretend fatigue became real fatigue.

The sweet scent of dying flowers wafts through the cell's window. Hunger makes my sense of smell acute. My body is in emergency mode; my mind jumps from one thing to the next. The only consistent thought is my yearning for hours to pass so I can eat again.

Is this really worth dying for? I tell the interrogators that I will not eat until I am out of solitary confinement. I am bluffing. I'd eat before I'd die. But how much longer should I fast? Shane, Sarah, and I promised one another not to eat for at least five days before the first hunger strike, but we didn't have an opportunity to discuss this one.

It is evening already. Perhaps I'll eat tomorrow night. I could drink my own urine like Gandhi . . . Years ago, I went without food for three days in the Sierra Nevadas with a bunch of friends. It was an inner cleanse and a meditative experience. Now, I just feel like a trembling, empty body, alone in the world.

Half-consciously, I stand up and move toward the door. The guard must have summoned me. He watches me grab the tray of shining white rice and a slab of kebab that has stared at me from the corner of my room since lunchtime. I dodder over to the trash can and pause for a moment. I hold out the food as if to give the guard one last chance to stop me. Then I let go, letting the meat and rice thump against the bottom of the garbage can.

I walk back to my cell. The guard sets dinner on my floor. The hot lentil soup smells delicious. The head guard appears briefly. "Shane and Sarah are nearby. If you eat," he says, "you can see them again."

Finally! The hunger strike worked again! I bolt my food. Even as I eat, I cannot wait for breakfast. I cannot wait to see Shane and Sarah. I shape my hair with my hand and some water, I tear the stray threads from my uniform, and I bide my time for the guard's return. A guard peeks in the window to ask, "Did you eat?" I show him the empty plastic bowl; then he leaves.

Hours drag on and he never returns. I realize I've been tricked. I bang on the metal door. I must see Shane and Sarah.

A guard turns off the lights for sleep, but I keep banging in the dark. I switch hands because my knuckles swell up. The metal clangs obnoxiously into the hallways. It must be almost midnight. Eventually, the head guard returns. Between the bars on the six-by-six-inch window in the door, he slips me *The Crusades Through Arab Eyes*, the book that Sarah brought on our hike. When I demand to see her and Shane, the guard makes an empty promise, then leaves me in my dark cell: *"Fardo."* Tomorrow.

15. SARAH

She must have heard me weeping from down the hall.

I'm curled up on the bathroom floor when the guard comes into my cell. The last week has been the longest of my life. Shane and I decided not to hunger-strike again. Shortly after our visit with Josh, my interrogators told me if we stop eating, it will delay the investigation, which they say will be over soon. We can only hope Josh, wherever he is, has come to the same conclusion. I sing loudly much of the day, especially when I hear the slap of sandals in the courtyard outside my window. I'm hoping the sandals might belong to Josh, and that he might hear me.

No guard has ever come inside my cell before. I look up at her through a cascade of blurry tears. I take her hand and she swiftly pulls me to my feet, wrapping her arms around me and holding me tightly to her chest for several minutes as I sob and wail.

I can't believe her kindness. She's young and beautiful. Her voice sparkles, bringing images to my mind of light scintillating the surface of a pond; of a soft, warm wind seducing branches to stretch and sway. She reminds me of a cartoon character, like Sleeping Beauty or Cinderella; it seems like the birds start singing when she comes to the door of my cell.

"I love you," she says to me in shy, broken English, putting her hands on my shoulders, looking me in the eyes, and raising her eyebrows like a question mark as her voice goes up an octave at the end of the sentence: "I love you?"

Her English has the wonderful quality to it of being a mere sound — she pushes out each syllable as if playing her mouth like a new instrument. Someone must have taught her the phrase and this is the first time she's had the opportunity to use it. Her eyes tell me she's waiting to see if the words will work.

Next, she helps me get dressed, helps me put on my sandals and blindfold, and leads me downstairs to the courtyard. This is where the guards take us twice a day to walk blind circles in the sunlight. All I can see out of the crack at the bottom of my blindfold is the area around my own two feet. For a few minutes I let my feet guide me, tracing figure eights in the concrete. My breathing slows and I begin to gather my thoughts. I don't understand how she does it. How does this young guard remain so cheerful in this dismal place? It's the first time I've broken down in weeks and she was there in an instant to rescue me.

Then she does the most miraculous thing. When I pass her in the courtyard, she begins stomping on the dry, dead leaves scattered across the yard. She smiles at me, inviting me to join in, to release some of my frustration, to play.

My legs feel heavy and weak, but I manage to make a few small, feeble stomps and smile back. When we get back to my cell, I hand her my blindfold and outer garments. I look into her eyes with gratitude.

"I love you too," I say.

16. JOSH

In my mind I am already running. My feet patter quickly on the brick floor. All day, my energy is dammed up, but in the courtyard, energy courses through me. They take me for two half-hour sessions per day. I'm allotted a single lane next to other blindfolded prisoners in nearby lanes. I jog back and forth blindfolded. It's the only time I feel alive all day — when I'm out here and thinking about escaping.

Escaping can't be as easy as it seems. I used to do the high jump in track and field, and I'm sure I can make it over this small wall. I peek under my blindfold while I skip to see over the wall. But I can't make anything out. I shouldn't run yet. I lob a date pit over it to hear the texture of the ground on the other side. It sounds like more asphalt.

I can't leave without Sarah and Shane, but I'm starting to worry I'll run on impulse. I worry that I will forget how crazy it is to think that freedom is beyond this short wall. Just because I don't see a guard in the normal lookout spot doesn't mean nobody's watching. First, I'm wearing a blindfold that I can barely see under. Second, there are windows looking out on the courtyard and probably cameras everywhere. I need to control myself.

There must be a way to escape this isolation. But the guards won't let me see Sarah and Shane. They've grown tired of my screaming into the hallway, "I want to see my friends!" They want me to stop pounding on my door. I don't know what else to do. The guards barely speak to me, even in Farsi. When they do, they hide behind the door and avoid seeing me.

I look forward to interrogations, but they come only once a week. The questioning forces me to reflect on my past. Why did I study environmental economics? Why did I leave my parents' home when I was eighteen? Why have I visited Israel? I get to justify why I went hiking, why I came to the Middle East, why I taught on a study-abroad program last spring, why I did community organizing in Oregon, and why I built cookstoves for villagers in Guatemala. These sessions remind me that I have ideals to stand by.

Otherwise, in the cell, the *blankness* is my enemy. I don't have a better word for it, but it's dulling my mind. It's a world where I can only reference myself in circular loops, where nothing makes sense, where

I feel guilty and worthless and think that everyone — friends, family, coworkers, lovers — must hate me and I can't do anything about it.

Memories sustain me. My starved mind conjures up images from my life to keep me going: the exhaustion of my cross-country bicycle ride before arriving at Niagara Falls; scoring three last-minute three-pointers to win the playoffs in the junior Jewish basketball league; marching twenty-five miles in the rain with a group of seventy-five protesters from small-town Oregon to our congressional representative's office to protest U.S. wars in the Middle East. Every image has a function: persistence, triumph, protest. I don't seek out these memories consciously, yet my mind knows where it needs to go to dredge up nourishment.

Every day, I recall a different year of my life as a way to remember who I am. I remember friends — Shane and Sarah, of course, and I think of my friends on the outside too. They'll wonder where I am. They'll worry when they don't hear from me for a while. Someone must know someone in Iran. I have a Persian ex-girlfriend. She'll do something. My brother. He's well connected and politically savvy. Alex will get me out. He'll notice I'm missing. He knows I went to Iraq. He told me that he loves me. I remember that. I wish I could tell him how much I love him. I wish I could see him. "Alex!" No response. "Alex!"

I have nothing to do in here. Having read and reread *The Crusades Through Arab Eyes,* I try to memorize the whole index — I have noticed that a page for Ibn Jubayr is mistakenly indexed for Jubail.

Once, when I heard a helicopter whirring near the prison, I deluded myself into believing freedom was imminent. I decided U.S. officials must be negotiating our release and that I'd be free within three days. I felt certain of this, though all I heard was a helicopter's blade.

Now I cling to the idea of being released on Day 30. In the corner of my cell, the corner most difficult to see from the entryway, there are a host of tally marks scratched into the wall. I check the mean, median, and mode of the data sample. The longest detentions last three or four months, but most markings are less than thirty days. I remember an Iranian American was recently detained and released from prison. How long was she held? Thirty days seems like a fair enough time for the interrogation to run its course and the political maneuvering to sort itself out. In my gut, Day 30 *feels* like the day we'll get released.

17. SARAH

"Shane," I cry out, knocking on his wall. We both jump up on our sinks and press our mouths to the vent. "Shane, do you hear that?"

"Yes, baby, it's terrible."

"It's a man screaming. It sounds like they're *killing* him. He is being *tortured*."

"I know, I know, it's so terrible." Shane sounds distraught. I feel my own eyes, wet with rage.

"It's not Josh. It can't be Josh."

"No, it's not. Baby, whatever they're doing to that man, they won't do it to us."

We stand there with our ears pressed to the vent, saying nothing, as the screaming continues. It's coming from outside my window, probably from the courtyard.

"I'm going to ring the bell and ask the guard what the *fuck is going on*," I say.

After a few minutes, a guard opens the door. It's the beautiful, young guard who told me she loved me last week. I immediately grab her hand—relieved it's her. I point toward the screams and then wrap my hands around my own neck, mock-strangling myself.

She looks back at me, concern in her face. *"Nah,* Sarah," she says, shaking her head. She raises her hand, palm out, motioning for me to wait, and closes the door. Moments later, she comes back with a smug smile on her face. She points in the direction the screams had come from and says, "Futball."

"What?" I know right away what she's getting at, but I don't want to believe it.

She points again and says, "Futball, Sarah, futball," a proud smile on her face. She's telling me that what I heard was sport, not torture.

I look at her with a mixture of horror and disgust. I get the impression she's repeating someone else's words, as though she just went and asked her boss what to tell me. I shake my head and she smiles back.

This is the woman who stomped on leaves to make me feel better. She held me and whispered sweet, unintelligible things into my ears. This woman is an accomplice to torture. There is a man being tortured outside this window, we both know it, yet she doesn't look concerned. She can even smile about it.

18. SHANE AND SARAH

Shane

Sarah and I are housed in the female section of the prison — I know that by now. Today, something is off. When we spoke through the vent earlier this evening, Sarah told me the female guard on duty seemed sick. When male guards bring me my dinner, they give food to Sarah too. They have never done that before — only women guards feed Sarah. When the men came upstairs this evening, I didn't even hear them coming as I normally do. Usually, they say *"Allah, Allah"* loudly as they walk up the steps to make sure the women have enough warning to cover themselves properly. This evening, they have been silent. The female guard must be gone.

After dinner, two men come around with a bucket of muskmelon. They open my little window and hand a slice of the fruit through. Then they insist I watch them pass Sarah's through the little slot at the bottom of her door. It's like they are making sure I know there is no foul play. Then, in their preoccupation with having me watch them feed Sarah, they forget to close my window.

My mind immediately leaps to what has been my preoccupation these days: escape. Recently, I realized that the little metal clips that hold the plumbing together at the back of the toilet could be used to pick the padlock that holds my window closed, not the tiny window on the door, but the one that lets in light from the outside. I don't know how to pick locks, but that's only a detail. I have plenty of time to figure it out. Never mind that reaching the window would be impossible unless I found something to stand on, which I'm sure I never will.

Sarah and I have talked about escape. Through our tube, after particularly difficult days or long interrogations, I've told her that I know where the guards hang the keys to the cells. I've told her that a guard once trusted me with a razor to shave with. I had to pass it back through a little window before he would let me out of the shower. I told her that usually the guards lock my cell door and take the key with them, but that sometimes they just leave the key in the lock.

I collect these details. Can there be any prisoner alive whose mind does not, voluntarily or involuntarily, collect and file the details that could one day lead to his or her escape? My plots and secret mental maps soothe me. When I imagine how Sarah and I would sneak down the hallways at night to find Josh's cell, I feel brave and strong. I need to always believe that, if I so choose, I — we — can become free the day that I just can't bear it any longer.

Today, the stars are aligned. When they gave me muskmelon, they left my door's window open. I look through the window and see the key is also in Sarah's door. There are no guards upstairs. If ever there was a time to act, this is it.

I reach my hand through the little six-by-six-inch window in the door of my cell, trying carefully not to touch its edges so as not to jostle the door. I reach all the way down to the door handle and feel around with my fingers. The key is there. I turn it slowly because I'm afraid that a click will echo through the silent hallway. My arm is barely long enough to pull the door handle down, but I reach as far as I can and when the bolt comes out of the latch, the door opens slowly against my body weight. My heart hammers in my chest. Will they hear it? I pull my arm back in slowly and carefully, stick my head out the door, and peek around the corner. A barred door separates our cellblock from the area where the guards normally sit. Somehow, I've never known that before. No one is there at the moment. The lights are off. It sounds like someone nearby is watching television. Maybe someone *is* up here. There are no cameras facing our cells. I pull myself back in and close the door quietly.

Sarah

I'm crouched in the corner of my cell, rocking slightly back and forth with my eyes closed, trying to remember the ten digits of my sister's cell phone number. I methodically punch them into the palm of my hand, waiting for it to ring, hoping she'll pick up. "Hey, Sar," she'd say in her soft, girlish way. Still, even in my mind, I can't really hear her voice. I can describe it, but I can't hear it. No matter how I try, all I can hear is the jagged, deafening whir of the fan between Shane's cell and my own.

This must be what happens; the more time I spend here, the more I'll lose the world. My memories will become mere shadows, with all the warmth and flesh drained out of them. Each of the twenty-one days I've been here feels longer than the last. I try to fill each minute, each hour, but with what? I think back to my life three weeks ago. There was never enough time then; I was always rushing from one thing to the next, chasing time. Here, time just sits heavy and solid like a giant boulder in my path. How can these minutes, these hours and days and weeks, really be objectively the same as the others I've experienced all my life?

My thoughts freeze at the sound of a door opening in the hallway. Is it Shane's? I sit frozen for several moments, listening. When I hear the door close, I sigh dramatically and try to unclench the muscles in my legs, back, and neck. It must have been a guard opening another cell door down the hall. I need to relax, I tell myself, or I won't be able to sleep tonight.

I hear a knock on the bathroom wall, climb up on the sink, and put my mouth to the tube. "Hey, baby, what's up?" I ask Shane.

"I have a question for you," he replies playfully.

"Go for it."

"If one of the nice guards offered to take you out into the streets of Tehran tonight, would you go?"

"Um . . . I don't know. It'd be risky. Still, if we got caught, it wouldn't be my fault. Sure, I'd do it."

"What if I found a way to unbolt the metal screens and crawl through the fan ducts into your cell, would you let me do it?"

"Of course I would."

I climb down from the sink to rest and stretch my neck for a minute or two. Shane and I now brazenly spend much of our time comforting and joking with each other through the plastic tube, but I can only talk in brief intervals before my neck begins to ache from holding it at a tense, ninety-degree angle. For at least ten days, nothing has happened—no interrogation, no news. We haven't seen Josh since we broke our hunger strike — it's been over a week since I've even heard his voice in the interrogation room.

Shane taps on the wall again. I get back up on the sink and press my ear to the tube. "What would you say if I told you I'd found another way to get into your cell?" he asks. "Would you let me?"

"Yes! Baby, this is kind of annoying. It's not fun teasing each other like this."

"Maybe I'm not teasing."

"Shane, what are you talking about?'

"Sarah, I can do it. The guards left the window open on my cell door, and they left the key in my door!"

"Shane, are you crazy? You can't just waltz out there!"

"I already did. A little while ago I stuck my arm through the window, turned the key, and walked out into the hallway. The TV in the guards' room is blaring, they can't hear a thing, and there's no one else out there."

"Baby, forget about it," I snap, both astounded and impressed that the cell door I heard open in the hallway a few minutes ago really was Shane's. "Forget it, no way, just stop talking about it!"

"Sarah, please listen. They always leave the key in your door. I won't make a sound except for the click of the lock; I'll be so quiet. The guards have already left. There's just one out there at night and they never check on us after dinner."

"If you're caught, they'll separate us, Shane. We'll lose everything."

"I won't get caught. Please, baby, let me do this. I need to see you."

I try to say no . . . I try to be strong and responsible, but there is no way I can resist the idea of Shane in my cell.

Shane

I crouch down near the bottom of the door and listen carefully to the silence for several minutes, trying to see if the guard has gone to sleep. As soon as I hear the tiniest sound, I get up and pace again. I pull up scenes from jailbreak movies in my mind, my only reference point for what I'm about to do. Like in the movies, I bunch up a pile of blankets and drape another blanket over them so they look like a sleeping person.

For hours I go back and forth from the grate on the door to lying on my back with wide-awake, gaping eyes. I think of Sarah, who I know is always attuned to the movements outside the cell at night in a way that, aside from tonight, I am not. She has told me she sleeps with the ceramic cover of her toilet under her head every night, ready to wield it against an intruder. We've both read accounts of women being raped while in custody here. And there is a pair of eyes that looks in on both of us sometimes. A man has stared at me as I sit nearly naked on the floor and scolded me whenever I met his gaze. When we saw each other after the hunger strike, I heard Sarah tell the translator about him. He chuckled condescendingly and retorted, "That is a woman guard. No men are going to look at you."

"But he has facial hair," she said.

"No men are going to look at you," he repeated, and sent her away.

I've told Sarah to knock on the wall when she can't sleep, but she never does. Is every night for her like this, hearing these little sounds that I'm hearing with my ear pressed against the door?

There have been no sounds for a while now. "It's time," I tell her through the tube. "Let's do it."

As I snake my arm down again, unlock the lock, and gently open the door, a screaming fear courses through me. I can't go back now.

Being out of my cell, closing and latching my own door, is like floating in purgatory, between my cell and Sarah's. It's better and worse than being in my cell all at the same time. It takes only one step to cross from the door of my cell to hers. This is taking forever. No, it is only taking seconds. Standing at Sarah's door, I am more exposed to the guards' station down the hall. My heart is ripping through my throat. I feel red. I'm trembling. I'm worried they will hear me breathe. I'm not even sure if I'm breathing. In one quiet smooth motion that sounds to me like a loud ringing clamor, I open the little window in her door to ensure that once I'm inside, I can reach through it and turn the handle to exit. Then, I go in. She is there. Her face is beaming and her feet are dancing nervously.

Sarah

I can't take my eyes off him. He turns around, closes the door, and gently closes the window with a little string he takes out of his shirt pocket. I watch him closely as he bends down and puts his ear to the slot at the bottom, listening for the gentle slap of dreaded footsteps. There are none.

Shane turns to me and our eyes meet. His eyes never had that quality before. He is undaunted by his own fear. This moment, like so many moments, feels surreal to me. At first, I'm watching it happen, like my eyes are trying to catch up to what my mind is telling me. Then, when Shane reaches out his hand to touch my face, it is suddenly happening to me and only me. Shane's breath is delicious. I look at his sweet face, his gentle eyes, and his sensuous, cherry red lips. My finger traces his lovely neck, strong shoulders, and dewy skin. His hands help me remember why I love having a body, not only a source of complaints and needs that I can't satisfy, but pleasure, beauty, joy!

I don't know how our clothes come off, but they do. Seconds later, we're on top of each other, around each other, and inside each other. What a joy to see Shane, who had only been a voice for me for three weeks, naked and alive, his face soft, his muscles tense, words of love and lust and longing spilling from his lips. I abandon myself. For fifteen or twenty minutes I forget everything else, the blindfolds, the interrogation chairs, the yelling, the screams, even the fear in Josh's voice as they led him away from us.

We have defied them; the fabric of this place is forever torn. No matter what happens to us in the next few days, weeks, months, these moments will live in me forever. I will carry this love like a shield.

Shane

It's dark. I don't hear the fans anymore, but they are spinning on as they always do. For once, the fans are our allies — they cover the sounds of two people starved for each other. I kiss her whole body softly. In this moment, the kisses on her smooth, radiant skin melt away my ever-present fear of punishment. We need to do this right. We don't know when we will ever be able to do it again. I feel electric inside, not like I did when I opened my cell door — that was an electricity of risk and danger. This electricity is the warm buzz of yearning, a current that knows it won't be ruptured, but will be nourished. When she lies on top of me, I am overwhelmed by the warmth of her body. It's only been a few weeks, but this feeling of another person's skin has been completely cut out of my life. Now her skin is all over mine. Her back arcs and she moans softly. The hard floor, the marble walls, the boundaries between each other; all are gone. The muscles in her thighs are mine. My hips are hers. She doesn't cry out like she usually does, but her deep gasping breaths make me crazy. In a flurry of breath and lips and skin and light we collapse together, my head in the crook of her neck and her hand on my back. We each say, "I love you."

Almost immediately, the fear returns. We fumble through our heap of clothes and get dressed. I give her one last, long kiss. We pause and look at each other; I squeeze her hand and go. There is less apprehension in the return trip because there is no other option. It simply needs to be done and it passes in a flash. I close her window and reenter my cell. I leave my window slightly open, my subtle way of mocking the guards.

Sarah

The next morning, the older guard is standing in my doorway. She's balancing a breakfast tray in one hand and propping the door open with the other. For the last week, I've been working hard to convince her to bring me two plastic cups of tea instead of one. Today, there is only one, and she looks at me apologetically.

I sit down and begin to butter the thin, flat, tasteless bread, adding two packages of honey and three dates I saved from dinner last night. As I sip my lukewarm tea, I watch the sunlight from the window casting shapes like little silver dancers across the walls. I had slept without fear or doubt. Despite the cold, hard floor of my cell, I felt like a woman who would wake up in the morning, throw on a robe and slippers, and water her plants while she brewed strong coffee and checked her e-mail.

The guards have already taken Shane out. I can hear his plastic sandals hit the loose tile outside my window every few seconds as he weaves in and out of the three lonely plants in the courtyard. I feel loved, deeply loved. I feel like I should feel. I walk into the bathroom and wedge my thumbs into the elastic band on each side of my pants, beginning to pull them down before I sit on the toilet. Suddenly, I notice something out of the ordinary and gasp. My mind flashes to

Shane out in the courtyard in his light blue prison uniform and I laugh out loud.

A few minutes later he's back in his cell and I'm standing up on the sink. "Good morning, Shane," I say cheerfully through the vent. "How do you feel?"

"Wonderful, I feel so in love with you."

"Me too, baby, it's incredible. Hey, have you noticed anything unusual?"

"Not really, what do you mean?" he asks.

"You're wearing my pants."

19. JOSH

There is still nothing to do — no new books, no communication with guards. They've even stopped interrogating me. Nothing. Stillness. My body aches from sleeping on the floor, and my soul took cover long ago.

The whirring of the fan drives me nuts. I take refuge in the bathroom because its door muffles the sound. I sit under the sink, waiting for the fan to shut off, knees to my chest, hands over my ears.

Day 30 has finally arrived, but I've braced for disappointment. I knew I'd based my hope on dreams instead of reality. I tried to stop believing my superstition that I'd be freed today.

Suddenly, the metal door rattles. I stand up from under the sink and emerge from the bathroom. A guard signals me to clean my room and gather my belongings. He must be releasing me. The floor is already immaculate — sweeping the floor with my hands is one of my favorite activities. I grab my book and three dried dates stuffed with pistachio nuts to share with Sarah and Shane. I wasn't crazy. Day 30 is for real.

I am in the hallway, blindfolded. Guards push me around and spin me in circles. I reassure myself that I'll have my dignity again soon. Freedom is just around the corner.

I arrive in an office; I sign and fingerprint paperwork verifying that they've returned the backpack and other things I carried on the hike. I enter the lobby to find Shane and Sarah holding hands. I dive into their arms. Relief pours over me in waves.

Just beyond the doors, a Peugeot waits for us on the pavement. We walk to it with our arms interlocked. Next to Shane and Sarah in the back seat, I give them each their date-pistachio snack. We cruise around alleyways at 15 mph. We encounter no traffic nor traffic lights nor pedestrians, only bureaucratic-looking buildings and occasional soldiers.

I can hardly control my joy. I turn to Shane and Sarah, hoping they will share my excitement. We start giggling — nervous laughter — at the comfort of our companionship, the absurdity of the hell that suddenly became a memory. Now that we're together again, the weeks of solitude I've just endured seem like a distant memory. Was it really a month long? Somehow this is funny to us, and laughter eases the tension.

Sarah tells me that she and Shane spoke to each other through a vent. *They what?* Sarah says, "I promise we didn't do it much." I can't believe they were near each other. *They had each other!* I had nothing. They also had a meal together without me. Why am I being singled out? These guys don't have a clue what I experienced. I would have done anything for a voice to talk to. I push the idea of them talking as far from my mind as possible, trying to convince myself of what I'd always assumed — we are in this together.

In the rearview mirror, I make eye contact with the stoic driver. He slows to a stop, then lifts the emergency brake. His gaze, knowing and pitiless, conveys the truth. This is not freedom. Shades and bars cover every window of the dirty, gray building before us. This is another prison.

"It's another step toward freedom," Shane says.

He said this several times during our first days shuttling around western Iran. I remember he said it on the way to the last prison. He said it when I first ended my hunger strike and had a meal with them. He must think it consoles me. But after a month of psychological torture, it doesn't console me at all.

New guards meet us as we exit the car with new, looser blindfolds — *cheshband,* one of the few words I learned last month. Sarah keeps telling the new guards that she needs to be in a cell next to Shane. We sign some paperwork. We hug and quickly exchange words of encouragement and solidarity. "Be strong," we tell one another, and, "I love you." Then a female guard unceremoniously ushers Sarah away.

A few moments later, the pace slows. Shane and I climb the stairs and proceed to a quiet hallway where a guard tells us to sit on the floor and wait. With our backs to a radiator on the wall, our knees to our chests, Shane asks me, "Josh, how you holding up?"

"Toughest month of my life. Trying to prepare for the long term. How are you?"

"It's fuckin' tough," Shane responds. "Did you hear the guy getting tortured?"

"Yeah, I was out in the courtyard at the time. It made me sick to my stomach. I had to stop exercising."

"Were you in the courtyard with the three sycamore trees?"

Suddenly, someone calls to us from down the hall. "Sssssssssss, do not talk." His diction is very clear, but he hushes us with the Persian "ssss" instead of an English "shhhh." I tilt my head back to see under my blindfold. The hiss came from a large, baby-faced man in glasses, sitting at a desk in the middle of the hallway. He then asks, "Would you like some food?"

"Yes please," I reply politely, hoping for an English-speaking ally.

He stands slowly and his footsteps approach lazily. He towers over Shane and me seated on the floor. We reach up like monks to accept flatbread rolled up and stuffed with a date-egg mixture. He tells us to call him Friend.

The bread tastes delicious compared to the dry matzo-like bread I was eating in the last prison. I relish the sweet filling. I close my eyes under my blindfold to enjoy it fully. Friend calls out from the desk down the hallway, "Do you want some more?"

"Yes," I say, excited to think he may in fact become a friend.

He delays a second before asking, "How does it feel to want?"

As if I hadn't learned that yet. As if I didn't already know that he has power over me. It was stupid of me to invest hope in a guard so quickly.

After an intake physical with a nurse, Friend puts me in a cell alone. I take stock of my new surroundings. In some ways, it feels similar to my last cell: same size and same kind of blankets to sleep on. Also, no doorknobs, no light switches, no bed, a small vent in the wall. But unlike my last cell, this one has no bathroom.

These details feel important. This place feels less well maintained. From the sink's pipes, plumber's tape hangs sloppily. Like the *cheshbands*, here, the carpet is looser. The doors have the same barred, six-by-six-inch window at face level, but here, they leave the window unlocked, allowing me to see the white wall across the corridor. The cell also has an English translation of the Quran. I open it.

"In the name of Allah, the Compassionate, the Merciful. All praise is due to Allah . . . Guide us to the right path. The path of whom thou has bestowed favors . . . not of those who have gone astray." When I read the words "gone astray," I think about the hike and put down the book.

Why did they transfer us here? Is this new prison a temporary holding place before release or a long-term place of detention? I need to know what's going on. I can't do another thirty days. I yell into the hallway, "English! I need to speak English with someone."

Several different guards show up, but nobody speaks English.

Finally, a swarthy, stocky man about my age stands before me expectantly. Unlike the other guards, he doesn't seem impatient or anxious. There is something soothing about his presence.

"Do you speak English? Where am I? What is happening here? What's your name?"

"This is Evin Prison. You arrived today," he says calmly. "My name is Ehsan."

"Ehsan. Yeah, I know I arrived today, but how long do people normally stay here?"

"Maybe one day." My heart jumps with excitement as he pauses. His eyes meander along the walls before coming back to mine. "Or maybe one year."

20. SARAH

She makes me take off all my clothes. The light from the hallway outlines her broad shoulders and stocky frame, all but filling the doorway.

"Off," she keeps saying in English, "off," pointing to my hijab, jacket, pants, socks, and underwear.

"Please," I beg her, "please help me. I don't know where I am."

"Off, *off!*" She's young, probably in her early thirties, but her face is stony and plain as paper, an almost gray color that matches her gray suit.

"Please," I say again, beginning to cry, "I'm scared. Where have they taken my friend and my husband?" I've taken to calling Shane my husband in prison, in the hope that it will increase our chances of being allowed to see each other. It's a habit I picked up while living in Syria, where most people don't recognize long-term dating as a legitimate tie.

Minutes ago, we thought we were being driven to freedom. When the car stopped in front of a new, much larger building, this woman, who told me her name was Nargess, immediately led me away from Shane and Josh to this cell.

"No talk, *off!*" she says again, locking her jaw and making her eyes bulge. She is extremely unattractive, I think. Maybe she got teased a lot when she was growing up and she likes this job because she can take out all her anger on scared, helpless people like me.

Still, right now, this woman is the only thing I have to cling to. She's also the first guard I've met who speaks any English at all. I have to find a way to make her like me.

I smile at her obsequiously and begin taking off my clothes. Even though at first glance she has few redeeming qualities, there must be something beautiful about her, I think, and I will find it.

And then I notice her eyes. They look like the ocean on a cloudy day, gray and green, deep and brooding. Her eyes are stunning.

"Your eyes," I say, pointing to my own, then to hers, "so beautiful."

For the first time she looks directly at me. The sides of her mouth twitch. Then, without warning, her face melts into an embarrassed smile.

"Yes," she says proudly, "I know." Quickly, she pulls her face back into a cold mask. Still, I caught her off-guard. Despite herself, she betrayed something.

"Off," she says again. "Off."

I'm now standing in front of this woman, completely naked. After working out every day in my cell for a month, any extra weight I'd gained from the ubiquitous Syrian *kinafe* (a sugary, cheese-filled pastry) was gone. Intense exercise was one of the only ways I'd found any relief over the last month — and I'm acutely aware of how sexy I must look.

Nargess points at the floor, motioning for me to get down on it. *Holy shit, this bitch is crazy,* I think, snapping back into the present. *Is she going to hurt me?* It suddenly dawns on me what she's asking me to do. She wants to see if I have anything stuffed up my ass or cunt. I hesitate a second longer, then mimic her obediently.

As degrading as it is, I feel oddly comforted by knowing this to be standard prison procedure. Satisfied by my squats, she hands me a pile of dark clothes and closes the door, leaving me standing naked and alone in an empty cell. I slowly pull on white cotton underwear so big that it almost slides off me, navy blue pants, a baggy T-shirt, and striped ankle socks. I set the blue plastic sandals neatly by the cell door, fold the small green towel, and arrange my toothbrush, toothpaste, bar of soap, and aluminum plate in a pile on the floor next to the sink.

There's nothing else to do, so I curl up in the corner and pull the coarse wool blanket around my shoulders. This cell is smaller than the last. The window is high up, covered in a sheet of perforated metal. I can't even see the sky. I close my eyes and feel fresh, warm tears running down my cheeks, washing them like a mother washes her baby.

As my ears adjust, a texture of sound gradually emerges from the silence. Somewhere in the distance, I hear the muted staccato of shoes on linoleum, the gentle crash of running water hitting a metal sink, and even the muffled sound of distant voices. I imagine rows and rows of cells spanning out in every direction, each one with a scared, lonely person inside. I hear a cough in the next cell and instantly think of Shane, remembering how he would cough to alert me every time he was taken out of his cell. I feel an emptiness inside my chest steadily expanding like a balloon slowly filling with water. After a few minutes, I muster the courage to cough back and my neighbor instantly replies in kind.

I'm still in the land of the living, I think, rocking myself back and forth. *There are so many others here.* I turn and run my fingers along the wall, pressing my cheek in the direction I imagine Shane and Josh to be. This wall, I think, is touching the floor on the other side of it. That floor is touching another wall connected to another row of cells. I don't know how many walls and cells are between us, but I know Shane and Josh are out there. This wall I'm touching touches them.

I see myself standing on a pile of boulders at the edge of the Mediterranean Sea in Beirut. I reach my hand out to Shane but can't touch him. The sea is as dark and brooding as Nargess's beautiful eyes. Its sound is deafening, like a thousand faucets crashing into a thousand metal sinks. Shane takes my hand and I realize I'm being pulled into a dream. Part of me holds back, wondering if it's really safe to dream, if it's even safe to sleep.

I will myself back into the room and force myself to open my eyes. Reluctantly, I crawl over to the small pile of things Nargess gave me. I pick up the metal plate and carefully prop it against the door. If someone comes in, at least I'll wake up.

21. SARAH

A few days later, one of the new guards, Leila, opens my cell door, and I spring into action. Leila seems to be the head guard in the women's section — all the others defer to her authority. She's a small, devoutly religious woman in her fifties. She and I can communicate a little better because she knows Arabic from her lifelong study of the Quran and I know a little Arabic from my studies in Damascus. She knows I'm innocent. The first time I met her, I told her I wasn't a spy. She said she knew that I had been in the mountains of Iraqi Kurdistan "for exercise" and that it was "problems between governments" that kept me here. Then, she took my plastic fork and gave me a steel one — a sign of trust.

I throw on my long-sleeved navy blue overshirt, or *manto*, buttoning it up frantically as I simultaneously slip on my hard, plastic sandals. I wrap my hijab around my head and tie my blindfold over my eyes. Lastly, I drape myself in a mountain of dark blue cloth, the chador, which I know from the early days after our capture is commonly worn by devout women throughout Iran. I carefully don each item of clothing until nothing but my face peeks out.

It feels like being shot out of a cannon. This is the first time I've been let out of my cell since they brought me here. I don't know where they're taking me, but movement feels incredible. My breath expands to fill the space around me, and walking down thirty feet of hallway blindfolded feels like charting a vast new territory. It's exciting to get even a sliver of a picture of where I am.

The guard leads me into a small room and my elation quickly evaporates. I take in my surroundings at a glance. The space is ten by ten feet, empty except for a desk and two chairs. There's a bare bulb hanging from the ceiling and a dirty, cement floor. The worst part is the foam. The walls, ceiling, and door are covered in thick foam. Soundproofing. The room is designed to muffle our screams.

They will torture me here. I consciously slow my breathing, feeling fiery ice shoot through my body as my heartbeat increases precipitously. I'll have to fight them, I think, as Leila closes the door behind me. Shane is no longer here to scream and fight beside me — I have to face this enemy alone. I'll go straight for the eyes. I'll use my fingernails to gouge out their eyes.

I look around the room again, searching for clues. I have read accounts of women being raped in Iranian prisons by their interrogators. This is at the forefront of my mind. Would they slowly build up to it, as a punishment if I didn't cooperate with their charades, or would it happen immediately as an effort to break me? "I'm made of steel," I mutter under my breath. "I'm made of steel and I can't be broken."

After about fifteen minutes, the door opens and a man walks in. My hands are white from gripping the sides of the chair and my jaw is clenched so tight, I can feel the blood in my brain. I'm guessing he's a tall man; all I can see of him under the edge of my blindfold are his legs and pelvis. Next to his beige slacks and black belt, his large, shiny dress shoes are so huge, they look almost clownish. He walks toward me and stops — his groin a foot or so from my face.

"Hi, Sarah," he says. His voice is young, with an intentionally forced and controlled civility to it. "How are you?"

"Very bad."

"So sorry," he says with no emotion in his voice. "Listen, I need you to tell us everything." He places a stack of lined paper and a pen on my desk. "Write it down, everything."

"I've already written everything at the last prison," I say. "Why can't you read that? There's nothing more to tell."

He takes a step closer to me. "Sarah, the other prison never happened. You need to write everything for us again, everything."

"Please step away from me." I clench my teeth and spit out each word like a small rock. *I'm made of steel.*

"Sarah." He laughs awkwardly, trying to sound unbothered. "Why are you so angry? Don't you want my help?"

"You will not come any closer to me than three feet," I yell. My voice sounds shrill. I fold and refold my hands on the desk. When I get angry, I fear they might betray my emotions, so I try to capture them.

"Sarah, listen to me. You have two paths, two options. You need to write every detail of your life. Do you want to stay here forever, for years? Or do you want to talk to your mother, to Shane and Josh, to go home?"

His voice rises and falls with practiced theatricality. "The truth will lead you to good things, Sarah. Lies and bad behavior will take you to *hell.*"

This is a game for him, I realize. He's actually enjoying this. He's clearly not driven by professionalism, or by a sense of right and wrong. He's here for sport.

"I need to see Shane and Josh," I blurt out. "I need to call my mother. I need a lawyer."

He takes a step closer to me. "Sarah . . ."

"Get *away* from me!" I yell, pushing my chair back. His voice and presence make my skin crawl.

He backs off. I hear him sigh and eventually leave the room. For several minutes I sit in silence, trying to regain my composure, terrified that I may have gone too far and really pissed him off.

Suddenly, a different voice speaks from behind me. "Sarah," it says, "no one is going to hurt you, I promise."

The voice is soft, gentle, disarming. I realize whoever it is must have been sitting there listening the entire time. "Who are you?" I ask. "How do I know you won't hurt me? I know what happens here."

"Believe me, Sarah, I will not let anyone hurt you."

"Why?"

"Because, I am assuming this is all a mistake, that you are innocent, but we need you to cooperate. The more you try and help us, the faster this will be over and you will go home."

"Why should I believe you?"

"Because it is my job to verify that you are telling the truth. Will you help me do that?"

"Yes," I say, "of course I will." I feel intense relief. This soft-spoken man is the opposite of that threatening oaf. My reaction to his kindness is as visceral as my terror was a few minutes earlier. I'm not going to be tortured or raped, I think. For the first time someone in here actually cares.

"Sarah, we just want to ask you some questions. Then, if everything is okay, you will go to court and the judge will send you home. Nothing will happen to you here, I promise." I immediately feel beholden to him. I take the paper from his hands and begin to write.

22. JOSH

The guard named Ehsan told me I could be here for a year. I better get my bearings. With no bathroom in my cell, I have to walk out in the hallway and pass other cells to get to the toilet and shower. Once, last week, I peeked into a cell occupied by two men in their twenties striding confidently around the room, vibrant and alive, arguing and debating. I hope they'll transfer me to a cell like that with roommates, vibrancy, life.

In this prison, guards don't hide their faces like they did in the last jail. Some even talk to me. One guard, who speaks a little English, taught me the Farsi word for the courtyard we go to, *hava khori*. He told me that it literally means *eating air*. I've even grown friendly with Friend, the guard who messed with me the day I transferred to this prison. I've treated him amiably and he's responded in kind. He speaks awkward English and tries out colloquial expressions on me.

He makes small talk, which can be the most significant event of my day. Friend gave me a bed and mattress, pistachios, bottled water, and crackers. He even gave me a small personal fridge that he put in the hallway in front of my cell. With snacks in front of me, I allowed myself to feel how hungry I've been, and how my stomach shrank after eleven days of fasting and four weeks on a prison diet. Relationships like this allow me to believe that though governments are often cruel, everyone has a soft spot, even my prison guards. Friend shows up at my cell to escort me down the hallway to the courtyard, *hava khori*.

"Do you know what a honey is?" He smiles goofily. "Do you have a honey?"

"No," I say dismissively, "and I don't want a honey in here."

I regret talking to him about English expressions. Why is he asking me about a honey? Why did he give me a bed last week? I push away the thought these questions raise.

At *hava khori*, he follows me in. No other guards follow me into the courtyard. Friend whispers; the rest of the guards speak in normal voices. Something is off about him, but I don't avoid him. I need companionship.

The courtyard is thirty feet long by thirty feet wide with twenty-foot walls. *Hava khori's* ceiling is the blue sky that floats beyond a grid of thin bars above; its eyes are the rotating security camera; its art is the patterns on the marble walls. Friend stays to chat.

I'd prefer if he'd chat with me during the twenty-three hours that I'm stuck in my cell than when I'm out here. But, I reason, human interaction is healthier for me than exercise. I tell him that I grew up near Philadelphia.

"I think that is near New York," he says. "I'd like to go to New York — I've only seen it in the movies."

I smile, thinking I too would like to go to New York.

"Why are you smiling?" he asks. "Do you want to show me around there?"

"If I get free, sure, I'll show you around." I imagine us walking around Times Square, going down to Ground Zero, dining at a Persian restaurant, and having a frank discussion about what it was like to be a prisoner and what it is like to be a guard.

"What do you think of 9/11?" he asks matter-of-factly.

"What do I *think* of it?" I repeat, perplexed. "It was bad. A lot of people died . . . What are you asking?"

He looks expectant, and I guess at what he is thinking.

"I don't think Bush did it," I tell him. "He might have known about it beforehand. We're trapped in a cycle of violence. September eleventh was blowback for U.S. operations in the Middle East. I think the CIA even armed Bin Laden in the past. I understand fundamentalists' anger, but the attack's not justified. Understanding and justification are very different things." I look around for a second. "*I'm* trapped in a cycle of violence."

He nods. He seems happy that I'm telling him what I think.

"It's just like my situation. My detention is blowback for decades of hostilities dating back to the CIA's coup in Iran in 1953. I know that. I get that. I understand why people would be pissed that the U.S. overthrew their first democratically elected leader. Just because the prime minister, Mossadegh, said Iranian oil should belong to the Iranians, not British Petroleum."

Friend says nothing. I'm not sure what he wants. Maybe I should sympathize and apologize for my government's actions. But would that mean that I'd be acting as if I do indeed represent my government? I remember when we entered Iraq, Sarah brought up the question of whether to apologize to the Iraqis for the invasion. I don't remember what I concluded then. I don't want to seem solicitous to Friend, but I want him to know that if I had been an Iranian thirty years ago, I would've been in the streets in 1979 with the revolution. I would have celebrated the ouster of a corrupt monarch who was controlled by the West.

I continue conveying to Friend how much I understand Iranian anger at the United States. "And the U.S. even armed Saddam to fight against Iran in the eighties. And the U.S. sanctions and pressures Iran about nuclear energy. The U.S. makes a big stink about Iran having nuclear *energy* while turning a blind eye to Israel's possession of nuclear weapons. I understand what your government is doing with me. But there is no way *this*"— I spread my arms out, indicating the high walls around me —"is justified."

I take a deep breath and realize how much I've let my guard down — how much I've trusted my enemy again.

"Wait," I say suddenly. "What do *you* think of 9/11?"

"It is not important," Friend replies abruptly and turns to leave. "You have twenty-five minutes left."

What an asshole! Why did I get suckered into treating him like a friend? Am I really that desperate for someone to talk to? Was that an informal interrogation? I can't believe I'm so stupid.

Friend stands at the door for a moment and calls out, "See you later, alligator." He won't close the door until I respond, humiliated. "After a while, crocodile."

23. SHANE

I walk into the interrogation room. My interrogator is sitting casually on the desk with one leg dangling over the other. Another tall man is standing beside him, but I can see only his hands and big feet through the blindfold. I sit down, silently, automatically, in the desk chair with my back facing them. It's been six weeks since our arrest. This is my third interrogation with this new, English-speaking interrogator. He questions me only. Sarah and Josh have different interrogators. I know this because while I am being questioned, I can usually hear their muffled voices nearby.

"Shane! Buddy! Talk to us. Relaaaaaax. We're friends, remember? Tell us what's on your mind." Weasel. That is my name for him. He's such a slick, slimy little fuck. I despise this man, but I need him. I want so badly to relax and talk. He knows that. He leaves me in that cell for days because he knows it will make me desperate. He's manipulating me. I can't allow it. I need to maintain a boundary between him and me. Or do I need to ease our relationship?

"What do you want to talk about?" I ask.

"Anything," he says. "*Sing* if you want to."

"Umm . . ."

"Shane, you know, I've been to America. I like Americans. Do you know why? They don't care about politics. In Iran, if you ask any shepherd about international politics, he will have an opinion. A strong one. Americans don't care. Americans are simple people.

"Okay, I have a question. But this is just as friends," he says.

He gets up and paces the room. He always does this when he is try-ing to sound smart. The soles of his shoes make loud, forceful clicks. "Shane," he says, pausing a moment for dramatic effect. "When was your first experience with politics? Your first memory."

I think for a while. "I grew up on a lake outside a little town called Onamia with a population of less than a thousand. About fifteen miles from the town was an Ojibwe reservation. There were a lot of Native Americans who went to my school. We even had a class called Indian Ed where we learned about native culture. Then one day, when I was in fifth grade, a bunch of white kids got up, left their classes, and walked out of the school.

"You know why they did it? They were demanding that the In-dian kids be banned from going to the public school — that they be forced to go to school on the reservation. I couldn't believe it. I had considered many of these kids my friends, but at that moment I real-ized they weren't. For the whole week, kids kept coming back in the morning and picketing out in front of the school. I guarantee you their parents put them up to it.

"At the end of the week, the administration gave in. A bunch of school buses came and picked up all the Indian kids and took them to the rez. They just drove the kids with brown skin away. The white kids all came back the next week and the picket ended, but the school canceled Indian Ed for good."

Are we bonding? Is he interested in me? Does he feel guilty for par-taking in my captivity?

"What about you?" I ask him. "What was your first experience in politics?"

"Shane, in our other sessions, I told you that I am the only one who asks the questions, remember? In our other sessions, you listed twenty-four countries that you have been to. Who funded those trips? How could you afford to go to all these places?"

I know what he is getting at, and it is a legitimate question. If I can't account for my funds, how can I prove that I am not being funded by the CIA? The problem is, I don't think my honest answer is that be-lievable.

"My family didn't have money when I was growing up. I remember a time when we didn't even have a telephone. My dad was a mechanic and when I was old enough, I learned to weld in his shop. I saved money working as a welder and when I was nineteen, I quit my job and traveled through Europe and the Middle East. When I came back, I started by living in a vehicle. It's called an RV and it's kind of like a van, except in the back there is a very small bedroom. By living in the RV, I didn't have to pay rent.

"Then, I started working as a journalist. I started using the money I was saving on rent to start reporting. I traveled to Chad and Sudan twice as a journalist, using my own money. I just kept writing and taking pictures. I wrote a lot about Sudan and published my photos wherever I could. A friend and I worked on a documentary about Darfur. I also photographed and wrote about poverty in the U.S. Then Sarah and I moved to the Middle East, partially so I could work steadily as a journalist."

Does this asshole believe a word I'm saying?

"Shane, you have been making good steps in our sessions, but you are not telling me everything. You need to help me help you. I am young like you, Shane. I want you to be free, like a bird . . . Tell me, did your government ever pay for any of these trips?"

Shit! He knows about the grant. I spent days worrying about this in the last prison, but it never came up. I thought they somehow missed it. I have to lie.

"No," I say.

He hands me a piece of paper. All pretense of camaraderie is gone; we are back in official mode.

Q: What was the name of the grant you used to study in Syria and Yemen?

A: I think it was called the Boren fellowship.

It came through the National Security Education Project. I applied for it in community college in 2003, a time when the government was practically throwing money at anyone who showed an interest in Arabic. The counselor at my community college encouraged me to take advantage of it, even though she knew I'd never want to work for the government. The grant stipulated that every recipient repay his grant with a year of governmental work, but my counselor told me not to sweat it. "None of my students over the past ten years have ever ful-

filled this," she said. If I was going to be a journalist in the Middle East, to monitor my government's actions, I needed to know Arabic. I couldn't afford to do it on my own, so I decided to go for it. I remember talking to Josh about it. Most of my friends thought it was a bad idea. Now, I think they were right.

Q: Who funded this grant?

A: The State Department.

"Are you sure it wasn't the Department of Defense?"

"Uhhh. I don't think so," I lie. "I think it was State."

"Are you sure?"

He is standing directly behind me, tapping his foot loudly.

"Yes."

Q: How did you enter Sudan?

Shit. He is laying out the parts of my life they have collected to make their case of espionage. My history of government funding and my history of illegally crossing borders. How did I enter? I drove through the desert, straight through a huge expanse of sand, across the border between Chad and Sudan. It was the only way for reporters to get into Darfur. The trick was just to stay out of government-controlled territory and to stay with the rebels.

A: I entered as a guest of the Sudanese Liberation Army.

He asks me about my trip to Baghdad, four months ago, when I was there reporting. How did you get a visa? Who did you work with? List every person you talked to in Baghdad and what you discussed with them. How did you get into the Green Zone? How did you meet with U.S. military? Who paid you? What were their religions?"

"Why do you ask me about Iraq so much?" I ask him. "We are in Iran."

"Shane, did you know that Iraq is part of Iran?"

"No. I didn't."

"We have always believed that Iraq is a part of our country. Iraq is mostly a Shia country. Did you know that Baghdad is a Persian name? Baghdad used to be a Persian city."

"Yeah, Persian and everything else. Baghdad was also Mongolian."

"Mongolian!? Do you know who converted the Mongols to Islam? The Persians!"

I don't comment on the fact that for the Mongols, assimilation was a military strategy, a way to ease their rule over foreign peoples. They converted to Islam to better control the Persians.

"Persian culture is very large. As an Iranian, I can go to Afghanistan, Uzbekistan, Bahrain, and many other places and feel at home."

"You know I've never met an Iraqi who likes Iran, Sunni or Shiite. Many of them think of you as occupiers, just like the U.S. They don't like your influence in their country. They don't like anyone's influence in their country."

"Yes. Yes. Iraqis are very nationalistic, but Iraq will again become part of Iran."

There is a moment of silence while the thought lingers between us. "Soooo . . . are you saying that Iran is going to retake Iraq?"

He pauses, then says, "Iran respects the sovereignty of all nations."

He changes the subject abruptly, switching from Weasel the lecturer to Weasel the interrogator.

"In November 2008, you received an e-mail of a dancing, flapping bird. Who sent this to you?"

"What? That was probably my grandma. She always sends me e-cards on holidays. It must have been a turkey for Thanksgiving."

Q: Who is Sarit?

A: She works for an Israeli human rights organization called Bet Selem. I interviewed her about weapons the Israelis were using against Palestinians in the West Bank — special tear gas canisters that were killing people and that almost killed my friend Tristan.

"Does Sarah know about this meeting?"

"I don't know. Probably not. She wasn't there."

"Don't worry," he says, as if we were having a moment of male bonding. "I won't tell her. You can trust me."

24. SARAH

"'Tiger, tiger burning bright.'"

"Excuse me?" I say. I'm back in the interrogation room, blindfolded, facing the wall.

"Two days ago, you wrote that you have a bachelor's in English literature. I'm trying to see if you are telling the truth. Do you know that poem?" I recognize the voice of the kind interrogator, the one who said he wanted to help me.

"Yes, it's William Blake."

"Good. 'Tiger, tiger, burning bright.'"

"'In the forest of the night.'"

"Good, Sarah. Very good."

"How do you know about English poetry?" I ask him.

"I also studied literature, mostly European but American too."

"Really? Where?"

"In Tehran."

"What are your favorite authors?" I ask.

"Virginia Woolf, Wordsworth, I like them all."

"Virginia Woolf is my favorite writer!"

"You have very good taste, Sarah."

His compliment fills me with gentle warmth and I can't help but smile. I love talking to this man; just being in his presence bumps my mood up a few notches. I'm still making a halfhearted effort to hold back, act tough, but in my gut I've already decided to trust him completely. Why shouldn't I? Right now, he's all I have, and I truly believe he wants to help me, to speed up this process and play his part in getting me out of here.

"Do you have any of her novels? Can you bring me one?"

"Which novel do you like?"

"All of them — *To the Lighthouse* might be my favorite."

"I will try."

25. JOSH

It's the second half of September. The light from the window hits the wall at a slightly different angle every day. I hear sandals shuffle down the hall toward the bathroom. They amble lethargically. It must be a prisoner. I rush to the window in my door. I clear my throat when he passes to grab his attention. The pale prisoner coughs back at me and he makes eye contact from under his blindfold. To see and be seen! I remember a Zulu phrase I learned in South Africa a few months ago, *"Umuntu ngumuntu ngabantu."* It means "A person is a person through other people."

My hallmates sometimes cough, or sometimes they whisper "Hello," *"Salaam,"* or "Be strong" on their way to the bathroom.

I am grateful for the eye contact with this prisoner, and I wait for him to return. My face presses against the barred window in my door; the fresh hallway air skims my face. I hear him flush the toilet.

This prisoner looks and even walks like a foreigner. From his cell down the hall, I often hear him thank the guards with a *"Merci"* when the meals arrive. How does he have so much gratitude? I've tried to mimic him and thank the guards for food, but I just can't bring myself to do it.

Walking back from the bathroom, this prisoner skittishly scans the hallway, then springs to my door.

"Are you one of the American hikers?" he says in fluent English. His gaze nervously flashes down the hallway to make sure a guard isn't watching. He tells me he's Belgian and was detained while on a bicycle trip when they came too close to a military site.

I answer as quickly as I can. "Hikers?" I stutter, "I — I — I was hiking in Iraq when . . ."

"I saw you on BBC. There are two others, right?"

"Yeah."

"They call you guys 'the hikers.' I gotta go."

As he rushes off, I realize the risk he's taken to talk with me, and I whisper, "Thank you."

26. SARAH

Talking to the gentle interrogator is the one thing that keeps me sane. We have real conversations, weaving from one topic to the next. One minute we're touching on our spiritual or religious views, the next we're delving into Iranian history before the revolution, American foreign policy, or my childhood. He has an almost fatherly energy about him. That's what I've started to call him in my head, Father Guy.

"Sarah," Father Guy says gently, "today I brought you two gifts — a new book and some photographs."

He places a book and several sheets of paper on the desk in front of me. They appear to have been printed off a website. At the top of the page it says, "Pictures of Hiker Sarah Shourd"; below is a series of photos I instantly recognize.

The first is a studio shot taken in the early 1980s of my mom, my sister, brother, and me when I was about five. We're wearing our best clothes, looking into the camera with big smiles and blow-dried hair. Then, there's one of my graduation ceremony at UC Berkeley and below it a beautiful shot of me climbing pyramids in the jungles of Chiapas in my early twenties.

The picture on the next page makes me stop cold. It shows my three young nephews belly-down in a muddy Georgia stream. It's a place we always go swimming when I visit. Somehow, I've avoided thinking about them up to this point, and now the feelings rush in. How will it affect them to have their aunt in an Iranian prison? What if I never come back? Will they become soldiers — want to kill Middle Easterners in revenge?

"What's the matter, Sarah? Aren't you happy to see pictures of your family?" I can hear the hurt in Father Guy's voice. I haven't thanked him for his gifts.

"Yes, of course, I'm just very worried about them."

"They are okay. Sarah, please, you must be patient," he says. "I'm trying to help you."

"It's just, my mother — I need to talk to her. This is unbelievably cruel."

Reflexively, I smooth the strands of hair that have come loose from my hijab and wipe the moisture from my forehead — not wanting to offend him. Every day, I pray that Father Guy will come to interrogate me. I sometimes spend hours rehearsing would-be conversations with him in my head. I've even become more conscious of my appearance; when I know I'm going to see him I pinch my cheeks and bite my lips to give them more color.

Since I was raised by a feminist single mom, my desire for approval from men is something I've always been a little ashamed of and tried to change about myself. I have no doubt it stems from growing up without my father in my life. In the case of Father Guy, I've abandoned my principles. He's the only person with any power over our situation that I can influence, and I'm determined to use what I have at my disposal — my femininity, my intellect, my sheer desperation — to make him like me.

As he paces the room, I try to imagine his face, which I have only gotten brief glimpses of through blurry vision from the gap beneath my blindfold. From what I can tell, he's surprisingly young, probably in his mid- or late thirties, with light skin and dark eyes. "Sarah," he says with concern in his voice, "you're like a sister to me. More than that, I consider you my friend. I know you're innocent and I want to help you, but what can I do?"

This is the conversation I'd been preparing for. "If you're convinced that we're innocent, then you have to help us," I say. "There is no religion, no moral framework, whether it be Christianity or Islam, that would condone taking revenge on the innocent for the crimes of their government. Helping us is your moral duty. You have no choice."

After almost two months in solitary, questions about morality and the nature of existence are constantly cycling through my mind. Even though I wasn't raised to follow any particular religion, a part of me has always sought a higher power. Most of the guards are devout Muslims — some even walk through the halls muttering passages from the Quran under their breath. Recently I've begun praying every time the call to prayer is broadcast over the loudspeakers. I've become consumed by a need for a belief system that can focus and guide my energy toward something much bigger and more important than myself. I need to see beyond my own pain.

"Sarah, you are right. When I think about your case, I feel guilty, ashamed. I want to help you. It is my duty to help you, but what can I do?"

"Have you spoken to your boss? Have you told him what you just told me?" I ask.

"Yes, well, it is difficult. I will speak to him, Sarah. I will do this."

"The thing I just don't understand is — why would God punish me? I've always tried to be a good person," I say. "Why am I here?"

"Sarah, I cannot know the mind of God, but I know you are a good person. I know that you help your students and you're brave to stand up to your government." He pauses. "Perhaps you are being tested."

"Thank you," I blurt out, then bow my head submissively. "Sir, forgive me, but —"

"What is it, Sarah? You can trust me."

"Well, I don't understand how you can work here, be a part of this. You're such a good man."

"I'm trying to do the right thing, Sarah."

"Well, perhaps it's not me that God's testing. Perhaps he's testing you."

27. JOSH

I find a pen full of ink hidden in my vent and my world changes instantly. With writing, my mind's floodgates open up. I can write letters. I can record my dreams. I keep a journal and plot my future. I write notes to Shane and Sarah. Don't people do their best writing in prison? Stories form out of thin air and I write them carefully, on tissues.

The written word has become my closest friend. Since recently receiving some new books, I enjoy allowing hours — sometimes whole afternoons — to disappear in fiction or poetry. My own writing on the tissues gaze at me, the blank space listens to me, and the words give me advice when I'm confused. The different voices of my mind can argue on paper instead of creating agony in my head. It makes me feel calmer, more collected, and less impulsive. I can think successive thoughts and remember where I started. I even feel my glimmerings of gratitude toward life.

I write questions to mull over. Would I kill a guard to get free? Would I kill two of them? What does being Jewish mean to me? How can I align my life with God? Can I love my enemy?

Since his informal interrogation of me at *hava khori,* Friend has become my enemy. Everything he does is a power game. It seems he's been setting me up since he offered me food the day I met him. He is nice for a bit, draws me in, then he burns me. This time, Friend stole my books, and books are my lifeline. I've been too friendly and too trusting with him. The written word is my only friend in here.

I play the scene over and over again in my mind, fuming: Minutes ago, he opened my door with books in his hand. But instead of giving them to me, he took both my books — *Animal Farm* and an ESL abridged version of *Pride and Prejudice.* He had asked to "see" my books, and I handed them to him — naively. He promised not to take them, but once they were in his hands, he closed the door and walked off.

I want to spit in his face. He once told me that he wanted to go to New York City. That is where I'll attack him. I'll push him into an alleyway, get him on the ground, and kick him and watch him bleed amidst garbage and rats. Bystanders will try to stop my rage. Then I'll explain to them that this guy took my books when I was in prison, and they'll cheer me on as I continue kicking him. I don't know if I've *ever* felt this violent before.

I ring the bell to call Friend. I don't know what I'll do when he comes. I need to get these books back even though I've read them multiple times already. I don't have enough words to fill my days and he takes what little I have. Every day, I tell myself that things are slowly improving — *slowly but steadily* — but losing two hundred pages is a huge step backward.

I ring the bell again. Anger is like a fire. It's burning me, but I want to burn him. I try to console myself: Maybe he will bring books tomorrow. Bullshit! No one responds to the bell I've rung. I am going to explode.

My anger proves his power over me. I need to find a way to calm down. I need to feel less impulsive. I hate myself for trusting him again.

Yet, I've believed that there is something good even in guards, that compassion can live even in prison, that I'm not surrounded by pure evil. I'm angry thinking that I've been wrong. I can see myself losing it and it's hard to calm down. I take a few breaths.

I take a few more breaths. I muster my compassion. I try to remember who I am and what I believe in. I need to touch Friend's heart. I'll try to understand him, work on our relationship instead of worrying about books. My only relationships are now with guards. I'll try to share my honest feelings and *request* the books — not even demand them.

Someone is coming. My heart races immediately. It is him. "What do you want, Josh?"

I look him in the eye for a full breath before I respond. "Friend, I'm confused. I'm confused, and I'm needing some clarity." I take a moment to let it sink in, a moment to let us connect. Then I continue. "When you said you wouldn't take my books, but then you walked off with them, I became angry. What was going on for you? Why did you take my books?"

Our eyes meet in silence; then he walks away, leaving me locked in my cell. Heartfelt communication doesn't work here. Nothing works here. The bars between us are too dehumanizing. Prison wins again. How am I going to survive emotionally with fewer books, one less guard to talk to, and a load of anger?

Twenty minutes later, my anger has burned down to sadness, and Friend returns. Before saying a word, he hands me a book between the bars of the little window in the door. Then he says, "Take this book. Josh, look, it has been a very difficult day for me. I'm sorry. My boss has been mean to me today, and I took it out on you. I'm truly sorry for this." He looks at me contritely and disappears down the hallway. I can hardly believe it. The book in my hand feels trivial compared to this triumph of love.

28. SARAH

"Sarah," Father Guy says, clearing his throat, "I am not supposed to be here. My workday is over — I should be on my way home to my family, but I couldn't stay away.

"I am sorry," he continues, "but what I'm about to say is not easy. I was wrong all these weeks when I told you your case would be treated fairly. Unfortunately, this is no longer true."

"Why?" I ask, not really believing him. "Has something happened?"

"Yes, I'm afraid so. I gave my report to the judge. I told him that there is no evidence of espionage; that you crossed our border accidentally. They have decided to put your case on hold. I came here to tell you that I might not see you again."

"What?" I'm trying to find enough air in my body to project each word. "What does that mean? When are we going to court?"

"I don't know, Sarah. It may not be for a long time and when you do go, it may not be good."

"But, you said . . ."

"I know what I said, but I'm here to tell you I was wrong. Your case has become political. Everything depends on your government now."

"What do they want, a prisoner exchange?" I feel my skin getting hot. No court means they don't give a damn about legality; they're holding us because they want something. "That means we're hostages." Uttered aloud for the first time, the word feels like a stone slowly sinking to the bottom of a deep well.

"Listen, Sarah, President Ahmadinejad is going to New York tomorrow, to the UN General Assembly. This may be a good sign. Something good may happen while he's there that will help you."

"Is he going to talk about us in New York?"

"Most likely, yes."

"Will you come back and tell me what happens?"

"Sarah, I may not be able to come back. Your case is closed. I'm not supposed to be here even now."

I can feel all the positive feelings I've been harboring for this man draining out of me as my body clenches. The last month has been a lie, a joke. The two of us have been playing at being friends, talking about God, reciting poetry, but now he's just going to walk away to safety and basically leave me here to die.

"Sarah, you know I've always tried to help you. I've done everything I can."

The worst thing about it is he's probably right. I was wrong to imbue him with so much influence and power — this is the guy who takes orders. He may hate this place, even hate himself for working here, but if he oversteps, he'll lose his job, and not even that will help me. There's nothing more he can do short of helping me escape.

"So, that's it? You leave me here to go crazy? That's the end?"

Tonight, he'll go home to his cozy family — I'll sleep in a cage. Even if they were leading me to the gallows, I'm now certain he'd do nothing. It's a good thing I'm blindfolded, I think, because I doubt he could bear to look me in the eyes. Still, I realize, the joke's on me. I actually care about him and I know I'll miss him.

"Sarah, I wish I could do more," he says. "It's wrong that you're here, but please remember that you are never alone. God is always with you."

29. JOSH

According to the tally marks I make on the wall, it is September 29 — almost two months since our capture. With each new mark I feel closer to being totally forgotten. It's late afternoon, my least favorite time of day.

To my surprise, the big-footed, oafish interrogator says we're meeting the Swiss ambassador. The Swiss represent the United States, which doesn't have an embassy in Iran. The interrogator escorts me outside the building along with Sarah and Shane.

The parking area feels like freedom. Without the high walls of *hava khori* to contain it, the sky seems vast. There's more than I can take in at once, and my eyes adjust to seeing distances. A majestic snowcapped mountain towers beyond the prison compound. Gorgeous foliage adorns the trees scattered around the building and up the mountainside. I'm shocked to see how hard September has worn on my friends, whom I haven't seen in a month.

Shane's eyes are set deep in his skull and Sarah's skin is way too pale. We walk around the prison compound and fill one another in on our interrogations — that they know about my family history and Shane's grant. Sarah mentions being disappointed that her interrogator is not here. I tell Shane that I'll pass my pen to him by hiding it in the garbage can at *hava khori*. We can all then pass notes to one another in the same trash can.

On a tall tripod, a video camera records us as we settle into a conference room. Sarah demands tea, and I'm amazed to see a guard hasten to boil water. The door swings open. Upon seeing us, the ambassador gives an audible sigh of relief. We stand up as she rushes forth for an embrace — her sky-blue hijab matches her radiant eyes and only partially covers her bright blond hair. The rest of the room fades to the background. I can only see her beaming presence. With her arms around me, she exclaims, "It is so good to see you! It is so good to see you!" She darts over to Shane and Sarah. We aren't yet seated when the ambassador introduces herself as Livia Leu Agosti and asks, "How are you? Are you okay?"

"Please remember," the oafish interrogator cuts in, "you have thirty minutes to talk. You are here — according to the UN statutes — to talk about health, nutrition, and safety." I look over to him, surprised to hear him use the word *please.* Looking at the oafish guy is the first time I actually see one of the guys from my interrogation. He's tall, oafish, and distracted. He has a thick black unibrow above jet-black eyes. I had thought all the interrogators were authority figures, but this oafish one is barely older than I am. A guard places three cups of tea on the coffee table. The video camera stays trained on us.

We sit and take turns answering Livia's questions. When she shares her perspective, I hang on every word. "You know, it is not exactly ideal timing to be here as an American . . ." In the summer, the largest social movement since 1979 shook the foundations of the regime. They were contesting the elections and the repercussions are still unfolding. I'm worried we may be perfectly framed as the generic

foreign agitators. They could say we tried to enter Iran to foment un-
rest. I wonder to what extent this complicates our case. The ambassador continues diplomatically. "Also, well, your government doesn't
exactly make things easy sometimes . . ." She trails off without being
specific and transitions to lighter topics. "Josh, many people like your
rap video. It was aired on CNN."

"No way! They aired that!" I cry out embarrassed over Shane's and
Sarah's laughter. "They better have shown our other music video with
Shane and Sarah dancing!" I join them to laugh like I haven't laughed
for months. I was so playful two months ago, rapping and dancing in
the towns of Iraqi Kurdistan. I like the idea of the world knowing I'm
playful, but if it is taken out of context, I worry that people won't take
my plight seriously.

We tell the ambassador we've not been interrogated for a while
and that we think they're done questioning us. Shane complains to
the ambassador that his interrogator promised that he and I would
room together when the interrogation ended. I turn to look at him,
surprised. This is the first I've heard of that possibility. My interrogator wouldn't utter anything besides the questions he had for me.

The ambassador takes it all in and her assistant avidly takes notes.
She gives us her business card and hands us letters from home, a few
books, and some Swiss chocolate. She offers to relay messages to our
families. My family loves Bob Dylan, so in my message I include the
quote, "Any day now, any day now, I shall be released."

As soon as we leave, Dumb Guy takes the business card, the books,
the chocolate, and the letters. The following week, Shane's interrogator hands me a package and says, "When you read these letters, you
should try not to get emotional."

In the cell alone, I grip the letters tightly as I read them, passionately holding on to every word. Tears flow from my eyes — tears of
laughter and sadness and joy. Blurry-eyed, I read the letters out loud
and wave them at the broken sky through my barred window. My
brother writes of playing basketball together. Mom and Dad write
with pure love. My friends quote Dylan and Rumi and tell me that if
anyone they knew would stay strong in prison in Iran, it would be me.
They had a vigil for us at UC Berkeley. I forgot how much they loved
me. I read long, sweet letters from my friend Jenny. I'd been hop-

ing she was thinking of me. I have been thinking of her — wishing I'd gone back to America to date her instead of visiting Shane and Sarah. I'd been meaning to date her for years — since we broke up in seventh grade. What have I been waiting for all these years? For the first time in almost two months, I realize I haven't been forgotten at all. For days, this knowledge consoles me and keeps me going through the agony of solitary confinement.

30. SHANE

To write about solitary confinement is to provide the texture of experience, but how do I provide the texture of an experience whose essential quality is its texturelessness? Solitary confinement is not a head banging against the wall in terror or rage. Sometimes it is, but mostly it's just the slow erasure of who you thought you were. You think you are still you, but you have no real way of knowing. How can you know if you have no one to reflect you back to yourself?

Would I know if I was going crazy? It all must happen imperceptibly in here, like a frog being boiled to death, unaware of the rising water temperature as he is cooked. The longer I am alone, the more my mind slows. I'm losing my capacity to think. I'm becoming an animal, just looking and feeling.

It's been almost two weeks since we saw the Swiss. I haven't seen Sarah or Josh since. The excitement of my family's letters has worn off — I practically have them memorized by now. All I want to do this autumn morning is to forget about everything and experience the slow daily birth from sleep to wakefulness, to sip my tea and eat my bread and jam. But I can't do it. I am unable to prevent my mind from being sharply focused on one task: forcing myself not to look at the wall behind me. I know that eventually, a tiny sliver of sunlight will spill in through the grated window and place a quarter-sized dot on the wall. It's ridiculous that I'm thinking about it this early. I've been awake only ten minutes and I should know it will be hours before it appears.

Sarah wrote a song in the first prison and she used to sing it to me. "You with your burnt green eyes, haven't you said to me, all they can take from us, is a piece of time?" I did say that to her, trying to say something that was both comforting and defiant. But I was wrong. They take everything from us — breezes, eye contact, human touch, the feeling of warm wet hands from washing a sink-load of dishes, the miracle of transforming thoughts onto paper with ink. They leave only the pause — those moments of waiting at bus stops, of cigarette breaks — they take those empty moments that we usually cherish amidst the fullness of our daily lives, and they shove them down our throats. They make time the object of our hatred.

Time here becomes different from anything I've ever known. We need events — and events are almost always interactions — to give shape to time. Here, time stops being something that moves me and everything else constantly forward from the past to the present, from the known to the unknown. It stops being a stream and becomes a shallow, fetid pool. I sit in it and wallow. I can't drain it and I can't move forward.

But somehow I always bear it. As torturous as the thought of the future feels, I can endure the present. No matter how tight the vise grips my chest, I always endure this minute. So that's what I do. I live from one minute to the next. And I try not to look at the light.

Eventually, I pull myself up and out of bed, determined to shake the ennui that always threatens to overtake me. I bend my step to one end of the cell, then back to the other. There are stretches of days where I live inside a poem by Whitman or Wordsworth or Ferdowsi, exploring and internalizing each stanza. I recently memorized Whitman's "To the Garden the World," so I recite it over and over again as I pace, usually jutting my finger into the air at the line: "Curious here behold my resurrection after slumber." My juices are starting to flow.

As I walk, my finger taps against my thigh in alternating short and long pulses, each accompanied by high-pitched vocal beeps. I recite the poem in Morse code, tapping out the lines, "By my side or back of me Eve following, / Or in front, and I following her just the same." I've been studying Morse code in the dictionary Sarah convinced the interrogators to give us by saying she wanted to study it like Malcolm X did when he was in prison. Now, we each take turns with it for a week at a time, passing it off to one another through our interrogators. Somehow the study of Morse code seems useful to me. It makes me feel like my life isn't seeping down the drain for nothing. Maybe I will use it someday. Maybe someday I'll be stranded somewhere with a flashlight and will be able to code my way to safety. Actually, I know that's bullshit. I know the only reason I study Morse code is because I need challenges like this to survive. To not do this would be to give up. To give up would be the beginning of the end.

I slide under my bed on my back and lift the end repeatedly as though it were a bench press. I do sit-ups and pushups. I jog in place on a stack of blankets and do high head kicks back and forth across the cell. I give myself a sponge bath in the sink and look at the wall again. The light is there now — a trickle of diagonal dots. The day has begun. I am hopeful that today my interrogator will come. Then, at least, I will have some human contact, an hour or two of conversation. As I wait, I sit on my bed and read the Quran aloud in Arabic for as long as I can take it. I can hear several other prisoners reading, creating a discordant harmony of chanting voices.

After lunch, the hours pass blankly until the light is on the long wall. The two large rectangles of ten vertical bars of light have fully gone around the corner. The day is at its midpoint. My afternoon depression is starting to sink in. My interrogator isn't coming. He never comes after lunch. All that's left of the day is stagnation. All I can do is wait for sleep. I've already juggled oranges and swept the floor clean with my hands. I do make one discovery: the color red is absent from my life, but if I close my eyes and put my face in the patch of sun, I can see it.

Then, I hear a distant voice. At least I think I hear it — that unique, high and guttural timbre of Ehsan. I press my ear to the window in the door and strain to hear it again. Was it him? I press the button on the wall. This man, the "officer" of the guards, is my savior. I've been expecting him to come for days. I want him to give me another one of the books the Swiss gave us. I finished my John Grisham novel too quickly — I couldn't help but devour it in a handful of hours. Whenever I have any books, and I only started getting books a couple weeks ago, I hide them in my cell because one guard sometimes takes them. Ehsan is the only person who will take a book of mine to Sarah's or Josh's cell, trade it with them, and bring a new one back to me. He is the only guard I trust.

Sometimes he comes by my cell to see how I am doing. Once,

he walked in on me playing solitaire with my illegal handmade playing cards and after asking me for them as if to take them away, he
paused and handed them back. We talk about his sociology studies
in college. I suspect he only stops by to let me vent, to make me feel
heard. He hangs his head and shakes it in shame as I tell him, vehe-
mently, how hard it is to live alone. Lately, my message is always the
same: "Why are we still in solitary? Our interrogations are over!" He
is genuinely concerned about this. I can see it in his eyes. He knows
we shouldn't be kept apart anymore. He has told me several times
now that he has tried to get in touch with our interrogators, but they
aren't responding. He tells me he is sorry.

I ask for him whenever I hear his voice, because he gives me hope.
He also challenges my old beliefs. I have been separating the world
into two clean-cut groups of people, perpetual enemies who are divided at the cell door. Ehsan makes his side a lot more complicated.

No one is responding to the bell. I pound on the door. Finally, a
guard comes. "Is Ehsan here?" I ask him.

"Nah," he spits, tossing his head disdainfully and marching off.

I don't totally believe him, but I try to forget about Ehsan. I close
my eyes while pacing and try to imagine myself strolling up my grand-
mother's driveway, walking through the market in Old Sana'a, or
walking down Telegraph Avenue in Oakland. In each place, I feel the
air and smell the smells. I look all around me and take everything in.

Then, still pacing, I start thinking of the books I want. The titles
comfort me: *The Brothers Karamazov, War and Peace, The People's
History of the United States.* I'm afraid that when the time comes to
tell somebody what books I want, I will forget one, so I recite them
occasionally to remember. I know that if that opportunity arises, I
will only have one chance. I've decided to keep a list of ten. Why ten?
I don't know.

Eventually I realize how quickly I'm bouncing from one end of
the cell to the next. I'm not walking so much as striding. And I'm
speaking out loud. I stop and look around me. How long have I been
doing that, repeating these titles out loud and counting them on my
fingers? How long? Something about realizing that I've been hearing my own voice — merely hearing it, not commanding it — fright-
ens me.

31. JOSH

I've tried to hide my religious and spiritual life. Once, a guard surprised me while I sat cross-legged, meditating. I felt awkward being seen, but he wasn't at all uncomfortable with me sitting like the Buddha. He told me I was lucky; he wished he had time to meditate like me. I hated him for saying I was lucky. But from then on, I became less self-conscious with my spiritual practices: meditating, doing yoga and qigong. Still, though, I don't feel comfortable with the guards' knowing I'm Jewish even though my interrogators already know. The more attention my Jewishness and my Israeli family gets, I reason, the harder it will be for the Iranian government to release me. I'm sure they don't want to risk looking soft on Israel.

A friendly guard takes me to *hava khori*. This guard taught me the Farsi numbers, the days of the week, and he even told me his name. He's tall, slim, and probably the youngest guard at twenty-three. Before he locks me in *hava khori*, he points to me and asks, *"Yahudi?"*

He is one of my favorite guards. I don't want to lie to him. Moreover, I'm tired of hiding who I am. I remind myself that being Jewish is not a crime, and that the interrogators already know.

"Yes." I point to myself. *"Yahudi."*

"Yahudi, no problem," he says, trying to assuage my nervousness. "Israel problem," he says. Then he leaves me alone to exercise.

I jog back and forth, questioning why I just told on myself. What if he mentions it casually to the other guards or his family? What if his father works in the media and leaks the story? My mind can't break free from these fears. I don't know if the media already knows that I'm Jewish, but I don't want to take even the slightest risk — not for a relationship with a guard.

He probably already knows I'm Jewish; otherwise, he wouldn't have asked.

Whatever! It was stupid. I should have played dumb, pretended to not understand him. I need to rescind my statement and sow doubt in his trusting mind. Thirty minutes pass at *hava khori* and the guard rattles the door open.

Before he takes me back to my cell, I urge him to pay attention to me. "Buddha," I say. I sit down in front of him cross-legged. *"Yehudi, nah.*" I point to myself, shaking my head. "Buddha." He looks confused and doesn't say anything.

As long as there are fair legal proceedings, my religion is nothing to hide. No Iranian official ever told me being Jewish was a crime. Iran has the largest Jewish population in the Middle East outside of Israel. Still, I fear how it'll be used for propaganda. He said "Israel problem" and my father is Israeli. Paranoid thoughts stir in my head. I think of Daniel Pearl killed on videotape by extremists in Pakistan after confessing his Jewishness. Stuck behind walls of fear, I can think of nothing else.

In prayer I catch a glimmer of mental freedom. I've scratched a few words — truth, justice, freedom, love — with the chalklike prayer ornament, a *turbah,* onto the surface of the southwestern wall, and I pray toward the wall five times a day or whenever I need solace. In prayer, I listen deeply. The silence of prayer is a good silence, a connected silence. Prayer calms me down. There is something about listening deeply that makes me feel momentarily free — free of the torments of my mind. Paradoxically, by listening to the silence, I feel understood by the world. No longer do I feel ashamed of who I am.

Prayer helps me believe that I still have control of my own fate: if I pray hard enough and true enough, I'll be released. It is the most active thing I can do. I kneel and stand, touching my head to the floor, imitating Islamic prayer when the call to prayer sounds over the loudspeakers. In my mind's eye I can see millions praying with me in unison, and that vision makes me feel less lonely.

Reading the Quran keeps me focused on the divine. I open to the Repentance Surah, which starts with three letters — three spiritual breaths: *"Alif. Lam. Meem."* The book provides no translation for these letters; they seem to me a spiritual invocation. I read the next line, "The Byzantines have been defeated in the nearest land." What does that have to do with *alif, lam, meem?*

My meditation teacher once advised, "Sometimes in a dish there are spices you don't like. Just remove that small black cardamom seed, then continue eating." I continue with the Quran. "They denied the signs of Allah . . . These criminals will be in despair." This is the part that my interrogators probably like. They pick and choose verses just like I do. Mercy is their cardamom seed.

I choose the Repentance Surah in the Quran today because it's Yom Kippur, the Jewish Day of Atonement. I mix religions, but it makes sense to me. All paths lead to the same source as long as I remove my cardamom seeds. Normally, I barely notice the Jewish holidays. Normally, my family pressures me to remember my religion. I'm guessing about the date for Yom Kippur, but I watched the moon whenever it angled through the window, and I calculated the lunar calendar.

Yom Kippur is a holiday for reflection, and I'm fasting like an observant Jew. It's a day to think over the past year and to atone for mistakes. Of course, I think about the hike: I should have taken responsibility and gotten a map. I shouldn't have blindly trusted Shane and Sarah. I think about my parents: I shouldn't have let politics interfere with our relationship. I think about how my brother quit school to devote himself to my freedom (a fact I learned from one of his letters). I remember being judgmental of him when I visited his friend's holiday party. All this repentance makes me feel like crap. It makes me feel blameworthy and that there is a good reason that I'm locked up. I wish I could call them and tell them how much I love them.

Practicing Judaism is a way to be closer to my family. I can keep kosher. I'll make the Sabbath holy: from Friday sundown to Saturday sundown I won't clean the floor, wash my plate, or even tally another day on my wall. I try to remember every holiday I can: Sukkot is coming up. I'll sing the *shema* — a prayer that starts with the word *listen*.

Suddenly, I'm feeling more Jewish than I've ever felt in my life.

32. SHANE

At the end of each hallway, there's a small open-air cell. The guards sat me in one of those today. I've been sitting here, listening to Sarah wail in a nearby room for a while now. "You have to let me call my mother!" she is shouting. "You can't do this!" Her cries sound desperate. I am boiling inside. I hate these people. What are they saying to make her wail like that? The interrogators haven't come to me yet, but these sounds coming from her are making me nervous. We were caught a week ago using illegal pens and secretly exchanging notes by leaving them at *hava khori* for one another to find. Ever since then, I've been waiting for them to come.

My interrogator enters the room. I know it is him because he stands silently for a while behind me, tapping his foot. He does this when he wants to make me nervous. "Shane," he says, beginning to pace, "what did I tell you when we first met?"

"I don't know. What?"

"That we were going to be — ? That we were going to be — ?"

I won't say it.

"Friends. I told you that I wanted us to be friends. But you blew it, Shane. You had your chance, but you blew it. You know what you have done is going to affect your case. The judge is going to consider the fact that you broke prison rules when he makes his decision about you." What a piece of shit. Our note passing is going to determine our fate? He can't even lie convincingly.

He slaps a piece of paper down on my desk. "Write," he says. Our interrogation has been over for a month now, but he knows the feeling that interrogation evokes. Intimidation. Fear. He knows that the very act of writing is punishment. He leaves.

Sarah's interrogator comes in. "How are you?" he asks, then adds, "I myself am not good." His voice gives an impression of an overworked, spent man, like he's been up all night, like he has stubble on his face and unkempt hair. "We trusted you," he says breathlessly. I hate his tone. It's as if he were hurt, like a father realizing his son has betrayed him. Fuck him.

Another interrogator comes in. He's the tall, oafish one with big feet who has been coming around more often lately, the one who took us to meet the Swiss over a month ago, the one I sometimes hear speaking to Sarah while she is being interrogated. He is always telling me not to worry about her, that she is doing wonderfully. "You are stupid, Shane," he says. "What did you think you were doing?" He comes in close behind me. "Stupid!" he says again, smacking me on the back of the head, not hard, just a humiliating slap that brings alive in me something reminiscent of all those school bullies who slapped, punched, kicked, and choked my younger, smaller, bespectacled self.

I jump out of my seat, pull my blindfold off, and spin around to face him. He is towering over me, at least a foot taller than I am. "Don't you *ever* touch me like that again," I say, pointing my finger toward his face. I don't remember the last time I've been swept away with such heat.

His jaw lowers and his square, pale face goes cold. He looks frightened, not frightened that I will hurt him — he's huge — but frightened that he made some mistake. The slap felt routine, like something he does all the time with people he interrogates. But we are high-value prisoners. He can't do this with us. He knows that, but until now, I don't think he knew that I knew that. I'm not quite sure why I do.

"Okay," he says, almost placatingly. "I won't do it again." I hold his gaze for a second longer than is comfortable, his eyes dripping remorse, mine full of fire.

Slowly and self-assuredly, I turn around, sit down, and pull my blindfold back down over my eyes. I pick up my pen and write: "I had a pen. I knew it was illegal, but I did it anyway. I needed to communicate with Sarah and Josh. It was a moment of weakness. It was a mistake and I will never do it again. If I do, I accept full punishment. I hope you find it in your heart to forgive us and to give us another chance."

I know that I have no choice but to return to my powerless role. The interrogators can always punish me by punishing Sarah and Josh; I take responsibility for the pen and bow my head because I want to take some of the heat off them.

Even so, something has shifted. I feel better than I did when I came in. I'm glad he slapped me.

33. SARAH

After the interrogators yelled at us for passing notes, I caught a brief glimpse of Josh and Shane in the hallway. Now, it's been two weeks with no sign of them. I have no idea when I will see them again — when I will see another human being other than the guards who hand me food three times a day.

Out at *hava khori*, I fall limp to my knees and press my forehead to the cold, stone floor the second I hear the steel door click shut behind me. The acrid taste of dust in my mouth makes me salivate as I begin to pray.

Ever since Father Guy left, the voices of uncertainty have gotten louder and louder. I have nothing left to cling to. Every time I start to panic, overcome by a gut impulse to fight or run, I bolt my feet to the ground and I force myself to pray for strength, for acceptance, for peace.

Still, I'm slowly coming undone. I've tried to forget Father Guy, and the hope he represented, but his voice is still in my head. "You are never alone, Sarah," he said to me on his last visit. "God is always with you."

Maybe he's right. What if God is with me inside these walls? If I embrace God, something so much bigger than my small, insignificant life, maybe I can let go of my futile attachments — to who I used to be, to a future over which I have no influence or control. Is God the answer?

Slowly, I drag myself to my feet, raising my gaze up to the cold, expressionless sun. "I surrender!" I whisper. The empty, blue sky glares back at me. "Do what you want with me, God!" I yell into the empty void. "I surrender!"

Back in my cell, the hours don't go any faster. I try to distract myself by studying Farsi script — which is almost the same as Arabic — on small scraps of paper and plastic wrappers I've collected from the trash at the end of the hallway. I sound out the words and memorize them, with the intention of trying them out with a guard later and maybe figuring out what they mean.

When the evening call to prayer finally echoes down our hallway, I turn on the faucet to wash my hands, head, and feet — the way Leila taught me — and kneel down to pray. Suddenly, I hear a familiar sound coming through my window. It must be coming from *hava khori*. I spring to my feet.

It's cautious at first, a slow, smooth whistle coming to me like honey in the air, like a scent or a color. My heart feels like a train in my chest — it feels like it could easily tear through my skin and spill out onto the floor. I leap up on the bed and stand on my tiptoes in the corner of my cell closest to the window, stretching my neck as high up as it will go.

"'Now I'm feeling so lonesome and I can't get you out of my mind,'" I hum to myself. I know this song. This is *our* song!

The tune is an ancient memory, Jolie Holland's "Sascha." Shane once sang it to me when we were in Beirut. After a night of dancing we were back at our hotel, my head on his lap, his fingers drifting through my hair. What happened to that relaxed, sensuous couple? I picture myself this morning in *hava khori*, yelling and flailing my arms in the air, speaking to God. What's happened to me?

When Shane was captured, he was wearing a silver necklace around his neck. The day I gave it to him, our apartment in Damascus was brimming with life — the music was blasting and people were passing around bottles of wine and *araq*, bowls of fruit and platters of pita bread, hummus, and pickled vegetables. Shane and I were ensconced on a small couch with two friends between us. I leaned over them to pass him a bottle of wine and put the necklace in his hand at the same time. He looked at it and gave me one of his huge, charmed smiles.

"There's a note inside," I said.

"What?"

"A note," I mouthed, and pantomimed pulling off the end of the silver capsule.

He knocked out the tiny scrap of paper and read it. "A part of me is yours forever." That's what I wrote. For one long moment our eyes locked across the smoke and din of the room. "I love you," he mouthed.

"I love you, Shane," I whisper. I close my eyes, let my head drop against the white wall of my cell, and imagine Shane's slender frame tracing circles outside in *hava khori*.

The memory of that moment in our loud, smoky apartment feels like another life — a life I've lost forever.

"I love you, Shane," I say out loud again. Something is coming to life in me.

The sound travels to me, through bars and night and lavender air, from deep in his body, briefly touching his lips. It circles my ears and dives into my soul like sweet love and pure emotion. With Shane so far away from me, it's sometimes hard to remember how things used to be. I've even questioned whether I made the right decision — moving to the Middle East, partnering up with someone so connected to this part of the world. Is all of this suffering worth it?

Shane's whistling is getting louder now, making his music splash like paint on these dreary walls. Something about the whistle reminds me of the old Shane — the man who's more passionate about life than anyone I've ever known. A few days ago, I was having chest pains, so Leila took me to the nurse to get checked out. I peered under my blindfold as I passed Shane's hallway, and there he was — stumbling blindfolded in my direction and carrying a small bag of trash. He looked so weak, so frail and submissive — so much like a prisoner. I inch higher up the wall, trying to erase that image of Shane from my mind; wanting to get closer to the beautiful man I love.

I'll never let this place come between us, I tell myself, letting my hands pass over my neck, remembering Shane and inhaling the sweet air that is him. No matter how long they keep Shane from me, he will be a part of me forever. Josh is a part of me now too. Everyone I've ever known is still a part of me — is still with me in here, and we're all a part of God. I close my eyes, a slight smile washing like a breeze across my face, and then the music stops.

The next morning the first thing I see when I open my eyes is a thin sliver of sunlight cutting across my cell. I must have slept in. I'm usually awake for hours, waiting for this slow march of light to begin. The stream is thick with dust motes, teaming inside it like fish in the belly of an ocean. As I stare at them, each gold speck of dust becomes a unique individual. That one is my mom, I think, and there are Shane and Josh. As I stare, each speck of dust becomes its own planet, then its own galaxy, and there I am, far below, crouched in the corner sitting on the floor in my cell. It's perhaps the most beautiful vision I've ever seen. What a universe, I think, amazed and strangely comforted. What a big, beautiful universe.

34. JOSH

Time is passing. The autumn chill makes me worry about the oncoming winter. Weeks crawl by with no interrogation, no Shane, no Sarah — nothing to make me think I'll be released.

A tall well-built guard charges down the hallway. I recognize him by his voice. He yells at my neighbor, who is returning from the bathroom. The prisoner, himself well over six feet tall, recently became the fourth person in the neighboring cell, which is also ten by four-teen feet. He barely wears his blindfold in the hallway, leaving it prac-tically up on his forehead. I've listened to guards yell at him about it all week, but he refuses to obey. I'm very impressed by his defiance.

The charging guard and the prisoner clash a few feet from my door. I hear a single smack. I rush to my door to listen as I cringe at the cruelty of this place. The prisoner releases a desperate yell. Then the drizzle turns to thunder. I hear each blow as they rain down on him just eight feet from where I stand. He screams as if being impaled with stakes. The whole prison must be able to hear him. A nearby inmate bangs on his door in solidarity. Almost immediately I chime in along with everyone else — all of us banging on our doors to protest the beating. The uprising is contagious, and the sense of rebellion has my blood rushing. This is our moment of power, our moment of instantaneous and blind solidarity. The guards can't shut us up, though they run frantically up and down the hallways trying. They seem scared and uncertain — emotions usually felt only by us prisoners.

We continue banging and yelling even after the beating stops.

The belligerent guard bursts into my cell. Fire rages in his eyes. His fists clench by his side. He's the one who did the beating and he's wound up like a bulldog on a leash. I take a few steps away from the door. He charges forward, fuming. Facing the beast eye to eye, I feel calmer and more alive than I have for weeks. "I DON'T WANT TO FIGHT!" I yell at him even though I know he doesn't understand English. He stares at me, deciding my fate.

He backpedals out of my cell as though yanked by a leash.

By late afternoon the sounds are long gone. The fight unleashed the violence that always simmers just below the surface. On my way to *hava khori*, I see the bulldog guard with his back against the wall, arms crossed and head bowed. He is now a contrite puppy. He raises his head slowly and gently takes hold of my arm. He enunciates slowly as if he's rehearsed the English words, "Excuse me."

35. SHANE

The old guard who smells like cigarettes is at my door. I am shirtless and a little flustered because he walked in on me exercising, running in place on top of a stack of blankets. I don't like it when this happens. I prefer to hear them coming so I can be mentally prepared, steel myself, and appear cool and unaffected by them. When I'm exercising, I'm in another world, running around Lake Merritt in Oakland or along the beach somewhere. He opened the door and snatched me out of it.

"*Jamkon*," he says, and waves his hand generally over my things. I've been through this before, and I hate it. He's telling me to get my stuff together. He's making me move.

I know the particular grief of cats when they switch homes. I know the nausea, the discomfort, the need to explore every corner and the many days it takes to settle in. I have switched cells five times in four months and I never know why they make me do it. Is it just to rob me of my little gains of stability? The cells are almost all identical, but when I have to move from one to the other, I feel uprooted. When I came to this cell, I panicked for the first time in two months. I was physically farther from Sarah's hall. The light was dimmer. The floor was covered in blankets instead of carpets.

But I quickly adjusted and appreciated the fact that this cell was farther from the guards' desk. It had a few plastic hooks on which I could hang my clothes. There was a little crevice I could use as a shelf. And the sink was out of view of the door, so I could sponge bathe in relative privacy. Now I don't want to leave this cell; it's my home.

The worst part about switching cells is that while I sit with my things neatly bundled up, waiting for the guard to return and take me away, I have to battle my own hope. I try as hard as I can to destroy the notion that we might be getting free. Maybe they are moving me to a cell with better heating, I reason. Now that winter has begun, the furnace in my wall has been blasting so hard it makes me sweat.

When the guard returns, he is smiling slightly, which I've never seen him do. I carry my bundle of blankets and he takes me to the door of a cell with a fridge — an object I know to be the privilege of only a few prisoners — sitting outside it. Can it be true? The guard opens the door and there is Josh, genuflecting with his head on the ground. He jolts up, looking stunned. "What's going on?" he says.

"It looks like I'm moving in," I reply. He leaps up, and we hug and laugh. The guard is now smiling widely. The cell door closes behind us.

This isn't the first we've seen of each other recently. For the past ten days, they've been allowing the three of us to meet for half-hour sessions at *hava khori*. These meetings have been mostly frantic, each of us desperately trying to unload what we've been storing in our minds for months.

During our first week in a cell together, Josh and I come back to life. The possibility of having a conversation on any topic for any length of time is overwhelming. We talk about Dostoevsky's *The Idiot* to an absurd extent, reading passages at random to discuss them as though they were Scripture. Josh gives me a lesson on the musical career of Bob Dylan and I school him on the Balkan Wars. Since we aren't allowed pens, I draw an invisible map with my finger on the wall. Josh tries to remember the Hebrew alphabet, which he learned in Hebrew school when he was nine. He teaches me the letters by writing them with sunflower seeds. The task becomes stressful because we have to destroy the letters every time we hear footsteps, lest we give the guards "evidence" that we are Israeli spies. We stay up late at night and discuss Josh's ideas about influencing the city government in Cottage Grove, Oregon. We make lemonade with the lemons from our lunch. We shoot hoops with a wad of paper and an empty box. We draw a ring on the floor and see who can toss the greater number of candies inside it. We tell each other "Good night," every night, before we go to sleep.

On our first day together, we come out to *hava khori* and see Sarah. "Guess what?" I say to her. "Josh and I are in the same cell now."

I can barely contain my excitement and for some reason I expect her to be excited too.

She takes a deep breath. "It's okay," she says. "I've been expecting this to happen. I'm not jealous."

The next day, Josh stays behind so I can be alone with Sarah for the first time in months. I stride into the courtyard as though I am bringing her a gift, expecting her to gasp in surprise to find us alone and throw her arms around me. Instead, when I pull my blindfold off, I see her at the opposite end of the courtyard, hunched over and staring at me with cold, angry eyes I barely recognize. I rush over to her and she steps back, like a cowering animal. My mind and heart are racing. What has happened? For every advance I make, she makes a retreat, always scowling. Then, her frigid eyes begin to tear. "It's not fair, Shane," she whimpers.

"I know, baby," I say, reaching my hand out. "It's not fair." I step toward her again.

As soon as I touch her, she starts screaming, kicking and punching the wall with each word she screams. "It's! Not! Fair!" Her explosion of violence clashes starkly with the stillness that surrounds us. The walls, the few leaves on the ground, the dust in the corners, none of them are stirred. No one bursts through the door to make sure we are safe. Nothing, except me, reacts to this terrible human eruption, which makes it at once pitiful and terrifying. I throw my arms around her, pull her away from the wall, and don't let her escape as she tries desperately to pull away from me.

"It's okay, baby," I say. In my mind I'm saying these words softly, trying to soothe away her enormous pain, but in fact I'm shouting, trying to get something through, competing with her screams and writhing body. What do these words mean when they are shot like a bullet out of my throat? "It's okay! Sarah. Stop! It's okay."

"It's okay?" she says sharply, looking at me like I've just smacked her. "It's not okay, Shane. This is *not* okay!" She's right, of course. It's not okay that she is alone and I am not. But I don't know what else to say. How can I possibly soothe her? Eventually, her rage shifts into sorrow, and something in her gives. She lets me hold her.

But it doesn't feel like she has found comfort in me. It feels like something is fundamentally broken, in me, in her, and between us. I feel like an accomplice in torture. Should I have refused to join Josh in his cell? But I didn't even know where they were taking me. I was never asked anything I could refuse. My body was just shipped around as it has always been, beyond my control or will.

But some questions start to gnaw at me: What if they'd asked me if I wanted to cell up with Josh? Would I have refused? If I could have, should I have? Every time I laugh or share a meal with Josh or stay awake longer than I would if I were alone, part of me feels like I'm turning my back on Sarah.

I remember when the Israelis bombarded Gaza at the end of 2008, while Sarah and I were living in Damascus. Suddenly, pictures of bloody, dead children popped up in bus stations around the city and all of the New Year celebrations in Damascus were canceled, from major concerts to small house parties. When I asked a Palestinian friend what he was doing for the holiday, he said, "How can we celebrate when people are being killed?" I couldn't really understand this at the time — didn't they want to forget about it, if just for a night? Now I get it. I know now that when people are completely powerless, the only solidarity they really have is in commiseration. But how do you keep yourself in misery and stay afloat at the same time?

A few days later, the three of us are sitting outside. Sarah is silent. Her jaw is clenched. Josh and I are on either side of her, sitting quietly, unsure what to say or do. Finally, she speaks.

"You could have refused."

36. SARAH

Fuck you, body, I think as I get out of bed. My head is a balloon filled with water, my shoulders are slack and lifeless, and my eyes feel like glass. I slowly pull on my pants, walk three steps to the sink, and drink two cups of water. My stomach makes an ugly sound, so I ring the bell for the second time even though I know it won't make the guards come any faster.

The "bell" is a round, black rubber button on the wall. When I push it, a green light, which the guards can see but often ignore, comes on outside my cell door. Fifteen minutes have passed since the first time I rang and I need to use the bathroom, so I begin to pace the length of my ten-by-fourteen-foot cell, punching the air like a boxer with my eyes fixed on my bare feet and the brown carpet. I let out a cry of frustration, pick up a plastic cup, and hurl it at the wall. I decide to pee in the sink.

Ever since I found out Shane and Josh were put together, I've been full of uncontrollable anger at everything and everyone. And hate — an almost violent hate.

I feel like I've really lost them. No matter what I do, I hurt. *If Shane and Josh can get through this, so can I.* That's been my motto since we came here. Even during the months we didn't see each other, I knew they were enduring the same empty hours I was. Their pain was my pain. Their hope was my hope. Now that they are together in one cell, there's a rupture between us, a distance I don't know how to bridge. I want to believe that their gain doesn't have to be a loss for me, but the truth is, as Shane and Josh become closer every day, I feel more and more alone.

When I'm with Shane and Josh in *hava khori,* I almost feel worse. Every touch reminds me of the absence of touch. Their situation seems heavenly to me — they're out of solitary! What could be better than sharing a leisurely game of chess, listening to endless stories about each other's lives, being able to connect without the fear of harsh, brutal interference? They are halfway there, halfway to sanity and normalcy, halfway to freedom! I want to feel happy for them, but the reality is that I don't know how much longer I can hold it together in this cell alone.

My time in my cell has become less and less structured as I get more depressed. I talk to myself, eat my food with my hands. Like an animal, I spend hours crouched by the slot at the bottom of my door listening for sounds. Sometimes I hear footsteps coming down the hall, race to the door, and realize they were imagined. Or flashing lights will dart across the periphery of my vision — but when I jerk my head around to see, they're gone.

These symptoms scare me. I'm certain solitary confinement is having an effect on my brain. Sometimes when I try to read, I can't focus and end up reading the same line again and again, finally hurling my book across the room in frustration. I've also become extremely paranoid about my stuff, afraid the guards will take things when I'm gone. I hide the food and other junk I hoard all over my cell—under the carpet, in my mattress—and check it compulsively. I jam the notes I've collected on scraps of paper and cardboard everywhere—inside the bedpost, under the sink, and inside a thin crevice on the back of the TV they gave me last week.

How will I know when I've left sane thought and behavior behind? When there's no turning back? Perhaps the biggest loss is my confidence. I've always clung to the certainty that I can emerge from this place unbroken and unchanged, but I'm not sure I believe that anymore.

I stare at the door, at its white, seamless face. I picture myself standing up, walking across the room, and ringing the bell. As soon as it opens a crack, I will throw my body against it. I will run down the hallway into the main corridor. I will run up to the first man in a suit that I see and I will wrap my hands around his neck and I will squeeze. I will squeeze his neck and look into his eyes as he tries to scream.

The fact that they put Shane and Josh together points toward hopelessness—it means we're not going anywhere any time soon. For months I have begged; I have pleaded. I've been kind and forgiving, but they've given me nothing. No phone call. No lawyer. No trial. Nowhere near enough books. No cellmate. I have been given nothing.

I take my beautiful, purple scarf and wrap it around my eyes. I wrap it around my mouth, gagging myself. I may be lost, I think, rocking back and forth with my eyes closed. I may be truly lost. *Stay with me*, I say to myself. *You're all I have. Don't leave me.*

The interrogators are fucking cowards. They know what they're doing to me, but they are too cowardly to come in here, to face me. Why am I being singled out for this torture? How can they leave me in here to go crazy alone?

Suddenly I'm on my feet, running to the door. I start banging on it with my fists, kicking it again and again. The guard opens the door and I stare at her, breathless and angry, my hands balled into fists. She forms her own face into a mask of steel.

"I want *hava khori*," I demand, my voice trembling, my face locked.

"No, Sarah!" she yells at me. "No *hava khori* today!" I hear the door slam. I hear her footsteps running down the hall. I don't hear anything else. I want to die. I want to disappear. I want to kill.

I hear a scream. It's far away, maybe in the courtyard or the next row of cells. There's something familiar, almost beautiful about it. The scream connects me to myself, roaring through my body and hollowing me out. Suddenly the door opens and a guard is in my cell. She looks at me with horror and through her eyes I see myself. The scream I heard wasn't coming from down the hall; it wasn't another prisoner in another cell. That horrible sound came from my own throat. It was me screaming.

The guard comes up behind me, grabs my shoulders, and begins to shake me. "Sarah, no! Sarah, *no!*" We fall to the floor, and I can feel her hands on my face, trying to get me to come back to my senses. I open my eyes and follow the guard's gaze behind me—where I see streaks of my own blood against the mottled white. I look down at my hands and begin to wail like a child. My knuckles are scraped from where I'd been beating them against the wall of my cell.

"I can't!" I yell at her. "I can't do this." Her arms encircle me now. "I can't!" I sob again, my voice quieter now.

My fists don't bring down the walls of Evin Prison, but they do cause a commotion. One guard after another comes to my cell, first cleaning the blood off the wall, then placing extra food beside me on the floor. When I'm still balled up and weeping hysterically after several hours, Leila comes in and helps me clean my face and get dressed.

Later that night, stern, beautiful-eyed Nargess comes to my cell door. "Telephone," she says, and rolls her eyes. She leads me to the pay phone mounted at the end of the next hall. I pass it every day, sometimes listening to other prisoners talk to their loved ones, but I have never been allowed to use it. I grab the receiver and press it to my ear, expecting to hear the soft, sweet voice of my mother.

"Sarah, we hear that today you are having a bad day," a man says in a smug, nasal voice that I instantly recognize as Shane's interrogator, Weasel, who told me he was assigned to Shane but had also come in to grill me from time to time.

"You can't just put Shane and Josh together and leave me alone," I yell at him. "I'm going crazy in here."

"Sarah, I'm calling because we are worried about your health. We do not want you to hurt yourself again, okay?"

"If you are worried about my health, then give me a cellmate."

"Sarah, you must understand, there are no suitable roommates for you here. None of the women speak English."

"I'll teach them," I say. "I'm an English teacher!"

I wish I could tell him about the clandestine conversations I'd had in English with other women in my corridor, about the prisoners who shout, "I love you, Sarah" down the hallway, whisper to me at night, or reach out to touch my hands through the bars as they pass by my cell. But I can't, so I say nothing.

"We think it may be dangerous for you, Sarah, to be with an Iranian. You are safer alone."

"*I am not safe!* You heard what I did to myself today. What if I get worse? Let me be with Shane and Josh during the day and then I'll go back to my cell to sleep at night." The real reason they don't want me to have a cellmate is probably because they are ashamed of what I'll learn about what they do to people here, the torture and rape; they know when I get out of here I'll have a huge platform from which to tell the world.

"Sarah, this is against the prison rules. Men and women must be separate. We gave you a TV—doesn't that help? Do you want Shane and Josh to be alone too? Why aren't you happy for them?"

"Don't you understand? *I can't do this.* I can't be alone any longer! I hurt myself today and I'll do it again. You have to help me."

As the words leave my lips, I realize how pointless they are. I've been begging and pleading and yelling for months and they've given me nothing. Nothing. The high-profile nature of our case protects us from being physically tortured, but the slow violence being done to my soul will leave no visible scars. *They don't care,* I tell myself. *Accept that.* Something inside me has snapped. I realize that I belong to this place—they own my time, my body, my future. No one is going to help me but me. I have to find a way to do this. I listen to Weasel's empty promises and hang up the phone. *It's up to me,* I think. *I have to survive.*

The next morning, Leila comes to my cell with a piece of equipment under her arm. She walks over to the TV on my floor and plugs it in.

"This is your new friend," she says to me in Arabic with a slight tone of sarcasm. I look down at the strange gray box. It's a DVD player.

37. SARAH

TV has become my world. Starved for every scrap of news I can get, I watch the English ticker on the Iranian state-run news channel as if it were a crystal ball. It was almost two weeks ago that Leila gave me a DVD player, and since then I've watched the one film I have, *Vantage Point,* seventeen times.

"Marg bar Amreeka! Marg bar Israel!" a crowd of angry, black-clad worshipers chants on the morning news.

The first time I saw footage of small crowds of conservatively dressed Iranians raising their fists in the air and chanting "Death to America," I freaked out. Now, these scenes don't bother me much. I've learned to twist and turn each headline in my mind like a Rubik's Cube, considering it from various angles in hopes of teasing out a kernel of truth. One day, I see a ticker that encapsulates all my deepest fears.

Sitting alone with my thoughts takes incredible discipline. Sometimes, I spend every waking hour between visits with Shane and Josh working through my rage. I know it's not Shane and Josh's fault I'm being kept alone, but sometimes it feels like they are flaunting what they have. An evil voice in my head tells me they don't really care about my suffering — that all human beings are essentially selfish and even the man I love is putting his sanity before mine. The rational part of me knows this isn't true, but at times my anger is all-encompassing and there is simply nowhere else to direct it. I pace my cell for hours, arguing with the walls. "They *do* care — shut up! I know Shane and Josh love me." I meditate, I pray, I focus all my energy on eradicating blame and jealousy from my heart.

Some days, it works. More and more often, I manage to walk out into the courtyard with a smile on my face. I spend our thirty minutes sharing my thoughts and small victories with them, and, since I'm the only one with a TV, I deliver daily updates on how quickly the world is going to hell.

"You guys, I have to say it: If the U.S. or Israel attacks Iran while we're in here, the most obvious way to retaliate is to kill us. They will kill us."

Josh lets out a slow, histrionic breath. "Sarah, I just can't believe that's a possibility," he says. "A big part of Obama's platform was that he was going to change his policy with Iran. An attack would make him look like such a hypocrite. And Israel won't act without the U.S.'s consent."

"Yeah, but Obama also said he was going to close Guantánamo," Shane retorts, "and that hasn't happened."

"The ticker said the U.S. is 'Amassing Warships in Iranian Water.' Warships? It doesn't get much more explicit than that."

Shane sighs. "Baby, let's not talk about this anymore—there's no point. Let's try to enjoy our time together." I let Shane put his arms around me and try to sink into his touch. These hypothetical conversations have become all too common. I know there's no point in thinking about the worst-case scenario, but that doesn't mean I have the power to stop either. For a few seconds, the three of us sit in silence.

"Sarah, for whatever it's worth, this is probably just a routine military drill," Josh says, taking my other hand in both of his and trying to comfort me. "Try not to take it too seriously, okay?"

One night about a week later, while arranging my blankets and getting ready to sleep, another ticker catches my attention: "Iranian in the U.S. Sentenced to Five Years in Prison." I do the math. I've already been here seven months. Five years would be sixty months. That means, if we're here for five years, for every day I've already spent here, there would be almost nine more like it. By the time I get out of this cage, I will no longer be me. I'll be a stranger.

The news didn't say his name or what he was convicted of, but what did that matter? Tit for tat, an eye for an eye. He gets five years; we get five years. Nationality apparently trumps guilt or innocence. Iran gets its revenge against the Great Satan, and the United States gets help discrediting the regime over its horrendous human rights record. Everybody's happy. Everybody wins.

38. SARAH

The guards in the hall are frantic. We're being led back to our cells from *hava khori*. I peek under my blindfold as we pass five or six young women lined up facing the wall. It's evident by their fitted jeans, long black jackets, and platform shoes that they've just been brought in straight from the streets. I sense their fear as I'm led past. "What's happening?" I ask the guards loudly. *"Chi shode?"* The women keep their heads down and say nothing.

Nargess presses her fingertips against my lower back, urging me to walk faster. I reflexively slow my pace, partially to piss her off but also to absorb as much as I can before I'm locked in my cell. When we arrive, she snaps at me angrily, slams the door, and stomps away. I quickly crouch down by the food slot on my hands and knees and listen to shouts and screams resounding off the walls.

It's December 27. I had a hunch something like this might be coming. I first got wind of it days ago on the news ticker at the bottom of my TV screen: "All Iranians Should Be United on Ashura." This relatively benign statement was followed by a thinly veiled threat: "Disruption by Hooligans Will Not Be Tolerated." The following day, I noticed two new cardboard boxes had arrived at the end of our hallway. After my shower that night, I hastily peeked inside and found dozens of light blue plastic sandals, the ones prisoners have to wear. *The guards*, I thought, *are preparing for an influx.*

Now, crouched by the slot in my cell door, I watch a procession of new inmates being led past my cell. One woman has a bandage wrapped around her head, caked in blood. Another limps past, her bright red hair streaming out of her torn headscarf, her head hanging low like a wilting flower. All day I've been desperate to know what's happening beyond these prison walls. Now the streets of Tehran are being brought to me.

The first ten days of the Muslim month of Muharram are a time of mourning for Shia Muslims. The last day, Ashura, is a major holiday—the day that Imam Hussein, the Prophet Muhammad's grandson, was killed at the Battle of Karbala in AD 680. Shia Muslims align themselves with Hussein, the brave underdog who stood up against the Umayyads, a much larger oppressor.

All month I've watched with fascination as crowds of men and women dressed in black ritualistically beat their chests on TV. Some use their hands; others use chains to self-flagellate. Neighborhoods organize crying circles, rooms full of grown men with fat tears wetting their thick beards, sobs escaping from their chests. After all these months of crying alone in my cell, public grieving seems cathartic to me. I watch the mourners and mimic their movements, genuflecting and gently beating my own chest.

I remember following Iran's post-elections protests on Al Jazeera back in our apartment in Damascus last June. Millions of people filled the streets of every major city, stretching beautiful green banners across city blocks. Then, when the government responded with violence, the color of hope and rebirth was stained with blood. By the time we were captured about six weeks later, it seemed like the worst was over. Maybe we were wrong.

After hours of commotion, the cells in my hallway are filled to capacity. Based on the number of women I've seen pass by, I estimate they have ten or twelve people packed tightly into each cell. I decide to ring for the guard and ask for my nightly shower. *"Nah,"* she says, exasperated, *"kaar daaram." I'm busy.* I begin to pantomime, performing like a trained monkey, smelling my armpits and crinkling up my nose. "Okay, enough, Sarah, quickly!" I grab my towel, hastily tie my blindfold around my head, and charge down the hall toward the showers.

I slam the bathroom door behind me and quickly begin undressing. I crank the hot-water knob as high as it will go, steaming up the room like a sauna. Is it really a revolution this time? If this government's overthrown, what will happen to us? Things could get really ugly before the opposition assumes power. We could get hurt, separated, or killed in the interim. I feel an affinity with these women—if I were Iranian, I have no doubt I would have been out in the streets today and I might have even ended up in here alongside them. But first and foremost I need to get home. I support this revolution—but there's no way I'm ready to die for it.

I suddenly hear the door open in the small room next to the showers. I hear voices—then the door shuts. Is someone out there? A barred window is usually kept unlocked so we can vent excess steam. I quietly unlatch it, peering into the small courtyard. A young woman stares back at me.

Think fast, I tell myself. It's been several months since one of the guards has made such a slip. I grab the bars between us, bringing my face as close to hers as I can, and begin to speak. "I Sarah. American. Long time here, no freedom," I whisper in my ridiculous, infantile Farsi. "You please phone mother Sarah. Sarah no spy, Sarah love Iran people, Sarah teacher, Damascus. Please you freedom phone mother Sarah, okay?"

The woman looks straight at me. "I know you, Sarah," she says in awkward but good English. "I am sorry, but I am not free, so I cannot help you." At that moment the door opens and Nargess starts yelling at both of us. She hands the prisoner a stack of navy blue clothes and ushers her away from me down the hallway.

As soon as I'm back in my cell, I hear a new prisoner in the solitary cell next to mine knocking on our shared wall. *Show some restraint,* I tell myself—mentally gluing myself to the floor. I want to shout words of comfort down the hall, but I have to be careful. I've already been caught twice—if I get caught a third time, there will be consequences. I can't risk losing my visits with Shane and Josh—I can't go back to twenty-four hours a day alone.

39. SHANE

Little cakes and candies wrapped in cellophane dash across the cell floor. The disruption stuns and freezes me momentarily as I kneel over tiny flash cards, each displaying a name, arrayed in a large network on the carpet. I was in the middle of something important: constructing an elaborate Greek family tree of gods and mortals. I'd been frustrated because I couldn't remember which marriages connected the House of Atreus to the House of Thebes. How many days must I study this before it finally sticks?

It takes me a moment to realize that the sweets were tossed in through the window in the door. Josh and I lunge to the floor, rip

open some cakes, and assume the position that new food always demands. We sit on the floor and each put a morsel into our mouths,
chewing it slowly with our eyes closed. It's spongy, like a Twinkie. The
concentrated blast of sugar is like an injection of well-being. I never
knew such a thing could make it through the walls of this prison.

After we finish, we set to work gently splitting open some dates,
removing the pit, and pressing some dark chocolate and a glob of
butter into each of them. This is our usual delicacy. We put six of
these in a small plastic bag that I then put in my pocket. Later, a guard
comes to take us to *hava khori*. Josh goes first, following the guard. I
trail in the rear and as soon as I get to the neighboring cell, I extend
my arm and jam the plastic bag through the little bars in one smooth
motion. Behind the door, I hear people scramble.

The cakes landed in our cell because of my curiosity about a question:

"Any news about the daughter of the sheikh?"

I heard this question repeatedly, usually whispered from the neighboring cell into the hallway shortly after the guards gave us lunch and
moved on to the next corridor. The question was always asked and
answered in Arabic, a language rarely heard in this prison. Judging by
their accents, I took our neighbor, the one asking the question, to be
a Saudi and the man who spoke through the vent across the hall to be
from Iraq or Kuwait.

Besides our visits with Sarah, eavesdropping on their conversations became the highlight of our days. It took me a week to gather
that the "sheikh" they referred to was Osama bin Laden, that his
daughter had been detained somewhere in Iran, and that she had miraculously escaped. I came to understand that the person who spoke
through the vent across the hall had a television, which was why he
was always the bearer of tidings, whereas our neighbor was always
the one with questions. Our neighbor always asked for updates on
the "Brotherhood," which I came to understand roughly meant al-
Qaeda, some other militant Sunni group, or all of them generally. He
asked often for updates on Yemen and Pakistan.

"The Brothers blew up some foreigners today in Pakistan."

"Hmmm," our neighbor mused aloud. "That will send a strong message."

Something about the way he spoke made it seem like he was playing a part, like he was trying to endear himself to his fellow Arab prisoner but wasn't himself a militant. In fact, he was sweet. Once he insisted on giving the man across the hall some money. He told the man he'd hide it under the cushion of the lone office chair that happened to be sitting in *hava khori* that week. It would be there, and he'd better take it. He needed to be able to buy juices and sweets from the canteen.

Sometimes, they talked about us.

"Hey, the other day when I went to the bathroom I looked in the cell of those Americans. It's like a five-star hotel in there. They have beds and a TV. And every day they go outside *twice*. What do they do when they leave? Pursue their hobbies? Yeah, Iran is *good* to the Americans."

I knew that last remark was just a jab at the regime, showing how hypocritical of them it was to be treating us so well, citizens of the Great Satan. So badly I wanted to interject, to tell them we didn't, in fact, have a television and that we weren't allowed to even make phone calls. And aside from all of that, we were innocent. But I resisted. I refused to talk, despite all of our neighbor's pounding on our wall and the subsequent coughs out into the hallway to get our attention. I refused because Sarah, Josh, and I made an agreement not to talk to anyone. The longer Sarah is alone, the more afraid — and paranoid — she has become that one of us will get caught doing something and that we will be separated from one another again. For her sake, we don't break the rules.

But now he is throwing cakes into our cell.

As the days pass, it becomes harder and harder to resist the urge to communicate. I am convinced it is safe — these two men talk to each other every day, loudly — and I am frustrated with Sarah's lack of faith in my judgment. How can we talk rationally about anything involving a level of risk, when conversations about risk become so emotionally charged? We can't, really, so I don't try. Instead, I will just give Sarah the news updates that I overhear in these men's conversations to supplement the scant pieces of information she gets from English-language programs on TV. The Yemeni government is crushing the Shia insurgency in the country's north. Bin Laden's daughter escapes from prison. An Iranian nuclear scientist is assassinated by a man on a motorcycle.

Eventually, Josh and I set a date: in two weeks, we will tell Sarah we want to actually talk to him. By then, Sarah's fear of getting caught should have worn off.

Then one day, at *hava khori*, Sarah says, "Guys, I really want to talk to my neighbor. I will be really careful. Do you think it would be okay?"

"Sure," I say. I am genuinely happy about this, because I want Sarah to connect with more people. What we have isn't enough. It isn't even a relationship anymore. It can't be, with just these thirty-minute snatches of each other. Sarah has been deteriorating. As the weeks pass, I feel more and more helpless. I try to think up topics of conversation. I psyche myself into fake shows of happiness to lift her spirits. But ultimately, nothing really works. She is becoming skittish. Her excitement at seeing us every day is desperate. She almost always leaves *hava khori* deflated and disappointed. People need people, and Sarah needs people more than most. It's as if they found a special little torture just for her, and put it on a screen for us to watch every day.

Yes, she needs to talk to her neighbors. She needs to break the rules. I want her to do whatever she can to get more contact.

"I'd really like to talk to our neighbor too," I say. "He seems to know a lot about what's going on."

"Okay, just be careful," she says.

Our neighbor says his name is Hamid, and when we start talking to him, our hall feels suddenly alive. Because of this one person, the fifth corridor is not afraid. We talk to him and he talks to the Kurd in his neighboring cell and the guy across the corridor. I start to notice little bundles of candy stuffed under the sink in the shared bathroom down the hall, waiting for some prisoner to pick them up. Sometimes, I see a pen hiding in the vent above the bathroom sink. People's incarceration dates and release dates start to appear on the wall. Inquiries into the whereabouts of specific people are scrawled in Arabic and Farsi onto the paint above the toilet too. Before people leave the bathroom, they pound on the wall behind the toilet, which separates the bathroom from our cell. By doing this, they let us know they are returning, so we can jump up and whisper *"Salaam"* as they walk by.

Josh and I start naming these prisoners who we see stroll by every day. We make it a game in which we imagine their lives.

"Did you see the new guy?"

"Which one?"

"The one who plays in an alternative rock band, who just wants to be an artist and left to play his guitar, and whose anguish is fueled by the fact that his government deems his music political and subversive?"

"No. I haven't seen him yet. I thought you were going to talk about the playwright — the one with white hair and stubble — who writes plays with subtle political undertones that satirize the Islamic Republic by outwardly condemning the shah."

The conversations with Hamid make our world exponentially bigger. He teaches us that we are in Section 209, Evin Prison's political ward. He teaches us the Farsi word for *hostage, gurugan,* which we use whenever we are frustrated with the guards, because they hate the idea that we would call ourselves hostages. And he becomes a conduit for information about our case through other prisoners, especially the three Iranians he shares a cell with. One of them tells us what the international news is saying about us whenever he has visits with his mom. He is an Iranian Marxist with a big, wild beard ("He's a communist," Hamid says, "but he's a really nice guy") and he has it hard. The regime hates leftists. In 1988, they executed thousands who wouldn't publicly recant historical materialism. This guy isn't an Islamist, but he says a change in government is the most important thing, so he stands with his Muslim comrades on the street to protest. When he came to prison, he says, his interrogators ripped out his hair by the fistful. He thinks he might get out soon, so Josh gives him his brother's e-mail, to tell everyone we are doing okay, and to wish him a happy birthday.

"Beware the big, quiet guard," Hamid says. "He can really lose it." He calls the guard Dog. We call him AK, short for Ass Kicker. He is the same one that burst into Josh's cell after beating a prisoner — the bulldog-looking one. He's the one who, not too long ago, went hall to hall trying to rally prisoners to chant *"Allahu akbar"* on what I took to be some special occasion. Those in Hamid's cell partook in the chanting, and snickered irreverently while doing so.

A week after Hamid and I start talking, Josh and I hear an argument between the Marxist and AK. All we can hear the Marxist say is *dastshuee, bathroom,* and the guard refusing. The Marxist begins to yell insistently and then we hear the distinct sound of flesh smacking against flesh. When it is over, we pound on the wall to get their attention.

"What happened?"

"The guard wouldn't let him go to the toilet. Then he hit him," Hamid says.

"Is he okay?"

"It's no problem," he says, almost chuckling. "This is normal. This is prison."

When Hamid and I first spoke, he was shocked that I could speak his language, Arabic. He said he knew our story, but he seemed circumspect in those early conversations, unsure whether or not to believe we weren't spies. But he was quickly set at ease, opening up as soon as I told him about my growing frustration with the fact that our government didn't seem to be doing anything for us.

He tries to reassure me: "America can do anything it wants to. You will be out soon. Trust me. Iran and America will do a prisoner exchange, then you'll be out."

I disagree. "I think the U.S. is just going to leave us here. We aren't worth much to our government. If we really were spies, we'd be out of here by now. Iran is going to keep making demands for things like prisoner swaps and the U.S. will refuse. Iran won't be able to back down. So how will we get out?"

I don't tell him that for months now, a feeling has been rising in me. It's something like patriotism, but that's not the word because it accompanies a growing anger at our government for not getting us out, for pursuing policies that leave so many people locked up or dead in this region, and for making the world more and more hostile to American civilians like us. No, the feeling is not patriotism ex-actly, but it is a kind of pride in being American. And it's not just that I miss home. The warm feeling I get when I read Whitman wax about the sweeping plains and gushing rivers is something different. This is pride in being from the home of Crazy Horse, Harriet Tubman, Frederick Douglass, Martin Luther King Jr., Malcolm X, and Nina Simone — people who really fought for freedom. It's a pride in know-ing that my friends — Americans — are fighting for us in a way that I know is transcending the juvenile discourse that got us into this situ-ation, where one country is good and the other evil.

The more I get to know Hamid, the more I see how similar his sit-uation is to ours, except that no one, not even his family, knows what is happening to him. Since he is Saudi, his government is one of Iran's main adversaries. He says he was arrested for a visa technicality, that he has never been allowed to call anybody, and that he has never seen a lawyer or his embassy. He just went to court and they gave him a one-year sentence for visa forgery. "But don't worry," he tells me. "Illegal entry is only a six-month sentence." We are coming up on five.

The relationship that emerges between our cell and Hamid's is one of mutual caring. We ask about each other's welfare so persis-tently that it's almost aggressive. He tosses things into our cell, like cakes and handmade playing cards.

This connection feels like the opposite of isolation. It feels strong and nurturing. But because it feels good, it also feels bad.

Immediately after I ate the cake, I felt guilty because I knew my en-joyment of it wasn't totally spontaneous. I thought for a second of how much Sarah would have enjoyed it in her lonesome isolation; then I suppressed it. Why shouldn't I enjoy something that is given to *me* for once? Why does every little thing that makes my life better have to make hers worse?

I was only trying to rationalize my greed. We could have shared the cakes with Sarah, but we chose not to. Josh and I didn't discuss it, but we were in agreement: Just shut up and eat.

Sarah would have shared them.

40. SARAH

I'm awakened to insistent knocking on my wall. The new prisoner in the cell next to mine hasn't stopped trying to get my attention since she came here after the Ashura protests. New prisoners in solitary are often desperate to communicate. They are scared and lonely, and they don't know yet how easy it is to get caught. If I knock back, it will only lead to more communication, like whispering into the hallway or passing notes. *There's nothing I can do to help her,* I tell myself sternly, trying to focus my attention back on my book.

Not even state-run TV has been able to censure the chaos and international outcry that followed the Ashura protests. The images they begin to show on TV are telling. One clip shows a group of protesters running from the police. Among them is a young woman with long dreadlocks streaming out of the back of her hijab. They show this clip again and again because it makes protesters look like unsavory foreigners. The government's desperate to blame outsiders — specifically from the West — for what's happening in Iran. I realize that doesn't exactly bode well for us.

"Sarah." I suddenly hear a soft whisper. The voice is close, almost as if it were in my room. "Sarah."

My head darts to the right and left, looking for the source of the sound. Am I imagining it?

"Sarah." The voice is louder now. "Please." It seems to be coming from the corner near the door, where my sink is. Above the sink is a vent. The *vent.* As soon as the thought crosses my mind, I leap off my bare mattress, compelled by a force beyond my control, and climb onto the sink. I press my mouth to the vent. "Who are you?" I ask the voice. "How do you know my name?"

"My name is Zahra, almost the same as yours, Sarah. I know you."

"You know me?"

"Yes, I saw your mother on TV. I am so sorry for you, Sarah. I am a mother too."

"Did you talk to my mother?" I almost yell, then remember to hush my voice. "Did you talk to her?" I whisper. As soon as the question escapes my mouth, I realize how irrational it is.

"No, Sarah. But I saw many pictures of you on BBC. You are a small, beautiful girl. I know it must be easy for you to be standing on the sink. For me, it is difficult. They beat me. They kicked me and tortured me. My hips hurt and it is difficult for me to stand." Her English is almost perfect, strongly accented with a sensuous, scratchy quality.

I feel tears welling up in my eyes. It's a miracle, I think. She knows me! We can *talk* to each other! How is it possible that the guards don't know about this? When Shane and I first discovered we could speak through the tube in the last prison, the same thoughts flashed through my mind. Is this a trick? Is it possible we're being recorded? Yet Shane and I were never caught.

"Is —" I hesitate, but I have to ask. "Is my mother okay?"

"I don't know, dear Sarah. I am Dutch," she says, "and Iranian. I live in the Netherlands, but my daughter is here, in Iran. They will not let me talk to my embassy. I don't know if my embassy understands that I am here. When you see your embassy, please tell them about me, Zahra Bahrami."

"They never let me see anyone," I tell her. "It's been three months since I saw the Swiss. I don't know what's happening. I've had no court, no trial. They won't let me see my lawyer."

"Yes, I know. They are liars, Sarah. Don't believe anything they say to you. I am your friend now, I love you."

I try to imagine her, hurt and alone, being taken out every day for beatings and interrogation and then put back in a cage. Separated from her children, her life, and her country. So much like me, I think. Yet, as bad as my situation is, Zahra's is worse.

"Why did they arrest you?" I ask.

"I was at the protest," she says, "Ashura."

Suddenly, the door of my cell bursts open. Leila's small, voluptuous silhouette is outlined in the doorway with her hands on her hips. She must have heard us talking and crept down the hallway — probably shoeless so we wouldn't hear her footsteps — and pounced on us in the middle of our forbidden conversation. It was like being caught masturbating by a ruthless schoolmarm — like discovering a private, secret pleasure and then being exposed and humiliated.

Not Leila, I think, of all people. I've managed to stay on her good side up until this point and she's helped me a lot — even complaining to the warden about how long they've kept me in solitary. Last week, Leila surprised me with a pair of purple cotton pants and a matching T-shirt — she knew my prison uniform was worn and full of holes. I can't afford to lose her.

This time my pleas have no effect. Leila's kind, motherly face slams shut like a steel door. She says she will tell my interrogators what I've done. Zahra, the new prisoner, is immediately transferred.

Now, when Leila comes to my cell, there are no more smiles, no conversations in Arabic. She hands me my food with a cold, unfocused stare, then wordlessly leads me out to the courtyard for a few minutes of sun. I'm no longer her *ukhtee al-aziza, her sweet sister* — I'm a plant that she gets paid to keep alive.

41. JOSH

It's early February — the anniversary of the Iranian Revolution. The authorities fear another uprising and they've emptied our hall of prisoners in preparation for snatching up dissidents. Hamid and his cellmates were transferred, and we were only able to sneak a quick goodbye. Flouting prison rules with Hamid and his crew made the last couple weeks the most enjoyable time yet. With them gone, and with Zahra moved away from Sarah, it's the end of an era of communication. Having more friends than just Shane and Sarah felt like another step away from solitary confinement. Now, I brace for the insularity of our triad.

Men and women are normally kept in separate hallways. However, our empty hall fills up with female prisoners. The Iranians treat us Americans as an exception to their strict gender segregation. I've seen so few women these past months: just Sarah and the Swiss ambassador. Recently, our captors gave Shane and me a TV, and I see some women there too. Now, Shane and I grow accustomed to women walking down our hall without wearing hijabs. One female guard with a moon-shaped face captures my attention. She has large black eyes, rosy cheeks, and an attractive inapproachability. I know I am desperately looking for signals, but I swear she sustains eye contact longer than is normal. As memories of romance drift deeper into the past, I idealize it more. When I joke with Sarah about playing matchmaker, she has a mixed reaction: "Josh! She is such a bitch!" Then Sarah realizes it's just a fantasy and says, "She *is* the cutest one."

The female prisoners on our hall are mostly in their twenties and thirties, and they are probably all activists. I don't think of them as fantasy objects like I do with the attractive guard. I feel solidarity with them. Torture seems allotted to them disproportionately. Half of the new female prisoners walk with a limp; one groans with each step. I'm hoping my new female hallmates will fill the void left by Hamid's departure. I'm not ready to accept my world shrinking to just Shane and Sarah again. Sarah says many Iranian women speak English. Maybe these new prisoners will have news we can't get from the tightly censored state TV.

I peek quickly into their cells on my way to *hava khori*. I see a woman alone pacing vigorously back and forth, counting something on her fingers. In another cell, I see a woman crouched in the corner, sobbing helplessly. It reminds me of my hellish first days of captivity. I want to reach out to them somehow, but the female guards patrol more vigilantly than the men. There's no time for more than a rushed *"Salaam."*

I tell Shane about the prisoners I see, but he's wary of my peeking into their cells. Guards often do that, and, in solitary confinement, peeking felt like a violation to me. Now, for us, it's a trifle. Shane reminds me what it was like in solitary and suggests that female prisoners are probably scared of being raped. If they see a man looking in, they may freak out. They may think I'm a guard.

I appreciate Shane's perspective, and I stop peeking. Indeed, I'm coming to appreciate him in many new ways the longer we cell together. We complement each other well. I teach him tai chi moves and he teaches me salsa steps. I love mulling over the cardboard pieces stuck in our radiator that serves as our chessboard and analyzing every possibility. Then Shane teaches me the obscure rules and the strategies he remembers from his grade-school chess club. If his lunch tray has more than mine, he gives me some of his food, and he expects the same from me. I love that he loves my sense of humor. I've never seen him cry before this, but often when letters from home arrive, he weeps, and I go over to his side of the room, hug him, and tell him that it'll all be okay. He does the same for me, and I know that he knows my pain. Our friendship feels deep, and it's getting deeper. I feel like I can tell him almost anything.

For our first several weeks rooming together, I continue to keep the Sabbath. We'd drink cherry juice and say some prayers over our dinner on Friday nights. For twenty-four hours I'd honor the day of rest: I wouldn't wash the dishes, exercise, or clean the floor. I realize that I've asserted my Judaism as a way to be the opposite of my jailers, but it's exactly this kind of binary thinking that created the political mess. Jews and Muslims aren't opposites. I don't want to think like that. We are both "People of the Book," as my guards often remind me. Such binary thinking leads to calling any American a spy, to suspecting all Muslims of being terrorists, and to ignoring complex identities, such as my own as an Arab Jewish American.

In these first months after Shane celled with me, I slowly let go of my religiosity. It's not that he challenges my practices overtly. His presence alone makes me question religion and puts me in touch with myself. Why am I pretending the Sabbath is ordained by God when I really don't believe that? I don't need to act Jewish as a reaction to their suspicion. With Shane in the cell, I don't need religion anymore. I don't need the Sabbath, nor do I need to drop to my knees during the call to prayer.

At *hava khori* I have to hide from Sarah the depth of my relationship with Shane. Sarah has said that she felt left out around Shane and me on the hike, and now the prison has institutionalized that arrangement. Shane and I try to avoid talking about funny moments in our cell or even the fact that the interrogators have let us have a plastic chair. Sarah asked us not to say *we,* but we don't always succeed, and she invariably shudders when we slip. I try to be sensitive to her, and I remind her that one day it will all be set straight: "When we're released, you and Shane will be together all the time. I'll be on my own."

Sarah and I very consciously build our friendship so our triad's dynamic doesn't all hinge on Shane. Shane encourages us in this. Sarah tells me about her youth as a punk and about her relationship with her mother. She suggests friends of hers whom I should date in San Francisco. I tell stories of my cross-country bicycle trip and about Jenny, the friend I've wanted to marry since we dated in middle school, how I received long letters from her in September, and how I finally feel ready for her.

Sarah truly opens up to me — in both her love and her anger. I've watched her rage and kick the walls. I've run in circles chasing her. I've tried to snap her out of her anger and was pushed away. We have cried together and hugged in moments of spontaneous joy. It's all very raw in *hava khori*. I feel so close to her, and I've taken to calling her my sister.

But even though we sometimes share such intensity, I sense her holding back with me at *hava khori*. Sometimes when things are hard for her, she just wants Shane to hold her. In these moments, she curls into a ball in Shane's arms. Other times when things are difficult for her or when she and Shane are arguing, Sarah wants me there for support.

None of this is articulated and I have a hard time guessing what's best for her. My offer to leave them alone in *hava khori* is the purest gift I've ever given. The hardest part is talking about it. Sarah seems to feel guilty for wanting privacy. I want what's best for them, even if it requires stepping back from the only two relationships I have.

One way we avoid the awkwardness of talking about it is by scheduling two weekly *hava khori* sessions in which they can be alone. On Saturday mornings and Wednesday nights, I go to *hava khori kucheek, small hava khori,* a room smaller than my cell but with a glass roof that allows me to see the sky.

During this time, I manage to fill my social needs. I sneak conversations with people taking showers in the room next door. I realize how desperate I am to talk to other people. I bring chocolate and a handmade deck of cards to give away. I develop friendships and look forward to meeting new people.

I meet a student who shows me his scars from being electrocuted, a Christian prosecuted for being a Muslim apostate, and a political science professor visiting from England for his mother's funeral. *Hava khori kucheek* is exactly what I need — exactly when I needed it.

42. SHANE

The censors neglected to remove a staple from our letters. I pull it out with my long fingernails and bend the end to make a little hook. I stick the hook delicately under the hem of my pink towel and pull the thread out, one stitch at a time. When I finish that, I do the same with my underwear, pulling out two long, white threads.

I tie the ends of these little threads to the zipper handle on my mattress. Like my sisters and I did as kids, I take the three threads and weave them together, tying little knots over and over again. The knots form into tiny little spirals.

When we were all in solitary, I read books like *Pride and Prejudice* and *Tess of the D'Urbervilles* and paced my cell. I would only read a few pages at a time before I would put the book down, pace, and think. What have I been doing all these years? I'd ask myself. Why haven't I proposed to Sarah yet? Are we going to just roll along together year by year, without ever deciding that it will be forever?

I turned over these questions for months. I've never really believed in marriage, so part of me wondered whether I was being seduced into it by my isolation and those nineteenth-century novels. Why trap yourself in an institution that is so feeble, that so often crumbles and wrecks lives? Why participate in the state's incentivization of one particular type of relationship: one man and one woman. I don't believe that it is somehow divinely ordained that humankind is meant to be monogamous, forever. It's so religious, marriage. And I don't really care for religion.

But I love her so much. I *do* want to be with her forever. I want to know that flame in her eyes will always ignite me and that the softness of her touch will always soothe me. I want to make a pact, build a foundation. I want to make something with her that is the opposite of this hell. I want to plan together. Though I see her every day now, she is still ripped away from me over and over again. The wound it leaves behind makes me double over with pain, a sharp, real pain in my stomach that lingers for hours. It leaves a dull, growing ache in my back. I want to be Sarah's sanctuary and I want her to be mine. There is nothing more important than this. I want to commit.

I would rather propose to her in freedom. I'd rather wait until it could be a true celebration. But the other day Sarah came out to *hava khori* with news. A ticker on the TV read "American Hikers to Be Tried Individually." Josh and I are sure this news means that Sarah will be tried and released before us. Releasing one of us might cool down international pressure on Iran and it makes sense that she would be the one to go. She is a woman. They would never release a man when they could release a woman. And Sarah doesn't fit readily into a spy mold. She is neither a journalist nor a Jew.

She doesn't think it will be her. She believes Josh will be the one. I think she's in denial. She might disappear from us any day now. I need to make sure she knows, before it's too late, that I want to be with her for the rest of my life.

I tie off the thread spirals and leave enough loose string at the ends to make sure I can tie them onto our fingers. It's date night, which means that Josh is going to stay in the cell while Sarah and I go out alone. I have the two little rings in my hand, white with a little strip of red in the middle, like a stone.

Josh has no idea what I'm about to do. I don't tell him because I'm afraid I might chicken out. I'm so nervous. What if she says no?

I get to *hava khori* before Sarah. It is dark and the late-winter air is a little cold. I lay a blanket down for us to sit on, under the camera so the guards won't be able to watch. Occasionally when we do this they come out and tell us to move into view of the camera. Usually though, they leave us alone.

When she comes out, I ask her if she wants to walk. I can tell she doesn't really, but I convince her to anyway. I want to loosen that constricted feeling that comes when I'm nervous. We do a few rounds, my heart pounding as I try to make small talk. Then I stop and sit us down on the blanket.

I take her hands in mine. We're both kneeling, sitting back against our feet. The single light, high up on the opposite wall, drowns out any view of the stars, but it casts a soft yellow glow on her face. Her hair is long now, drawn back in a ponytail, but I can't see much of it under her purple hijab. Her lips are slightly redder than usual, probably from the strawberry jelly she uses sometimes for lipstick. She is beautiful.

It's quiet out. No one is around. It's just she and I. I am aware that this is one of those moments on which life pivots. I can see in her eyes that she doesn't know what I'm going to say. I'm shivering slightly even though it isn't cold out.

"Baby, I didn't want to do this here. I wanted to be somewhere beautiful, but —" She looks confused and a little worried. "Will you marry me?"

Her body jolts with surprise. She squeezes my fingers. Her smile — my favorite part of her body — beams from somewhere deep. She sits upright. For a few moments, she says nothing. I hold my breath. Then she says yes.

I tie the rings onto each of our fingers and we hold each other, looking into each other's eyes, smiling.

43. SARAH

The next day, Leila brings me to *hava khori* and gently closes the door behind me. Stepping into the evening light, I close my eyes and take in a slow, deep breath. The cold air has a faint warmth behind it — a sweetness I can almost taste. My ears even detect the whisper of a breeze coming from somewhere behind these high walls.

There must be flowers out there, I think. After all, it's almost spring. I open my eyes when I hear the click of the steel door opening behind me. Shane and Josh are taking off their blindfolds. I smile when I see they are with our favorite guard.

"Ehsan," I say as I approach them, "did you hear that Shane and I are getting married?"

"Yes, I know, I am very happy for you." He smiles, leaning on the door frame of *hava khori* with his arms crossed. Ehsan is an officer, a step above the other guards. He's clearly uncomfortable with our imprisonment, and from day one he's tried to help us. When we were all still in solitary, Ehsan would come to our cells one by one and offer to rotate our reading materials. Again and again he's complained to the warden about my solitary confinement, trying to come up with alternatives to keeping me alone.

"I wish you could come to our wedding," I say. "It's going to be so beautiful."

"*Inshallah*," he replies, "it will be very soon. How much time do you need together today?"

"As long as possible," I say. "Maybe two hours?"

"Yes, I will see if I can do it."

"Also," Shane asks, "can you get us more yogurt and cheese? The other guards have been refusing to give us extra."

"Yes, of course," he says. "I will go to the kitchen. Is there anything else you need?"

My heart quickens at the prospect of spending two hours together—I haven't had that much time with other human beings since interrogation. That much time with Shane and Josh will raise my spirits for days. My anxiety level will go down, making it easier to read, study, and focus during my time alone.

I make a point to hug Josh first, holding on to him long enough to feel our bodies relax into each other. The guards always tease Josh and me for hugging, carefully explaining to us that men and women don't hug in Iran unless they're married or family. But we are family.

"They let us shave yesterday," Josh says, sitting down on our gray wool blanket and tucking his legs to one side underneath him. "When I saw my face in the mirror, I was startled. I look harder than I used to, more like a prisoner. If I didn't know me, I'd say I looked guilty."

In some ways Josh reminds me of Ehsan. Just like Ehsan will never don the callous shell of a prison guard, Josh will never take on the persona of a prisoner. Even after six months in here, he refuses to wear his blindfold properly.

"No, Josh," I say adamantly, taking his hand, "I wish you could see what I see. Your face could never look hard." You can dress Josh up in a prison uniform, blindfold him, leach out all of the color from his skin, and swell up his biceps from incessant exercise, but you can never take away Josh's beautiful, electric smile, or change the sweet essence of the man he is.

It seems to me the longer that Shane and Josh are kept here in these shadows, the brighter and more beautiful they become. In many ways this place brings out the best in us. Shane and I pass a slightly softer blanket back and forth every day, taking turns immersing ourselves in each other's smells by cuddling with it every night. Josh gives me his favorite tank top because my shirt is full of holes. "You will not obtain righteousness until you give from what you love," we remind each other, quoting the Quran.

I ask Shane and Josh if they'd like to hear a new song I've written. They both lie on their backs, folding their hands behind their heads and closing their eyes — the optimal position, they agree, for listening to songs.

"The promise of freedom," I sing, "and the sound of your voice."

I let my voice trail off and listen to its echo from the high walls. When I sing to Shane and Josh, I'm also thinking about all the other prisoners sitting in their cells; I hope they can hear me. My new song is about my mother's voice — which is beautiful, deep, and strong. It was my mom who taught me how to sing. Growing up, I always wrote songs, but I had never taken them seriously till now — I'd never really needed them. In prison, singing has brought me back from the edge hundreds of times.

"I learned an amazing word," I tell them when I'm finished. "Egress."

"That sounds familiar. What does it mean?" Shane asks.

"It means freedom to leave whenever you want," I say.

"The promise of egress," Josh says, "like your song. That's exactly what we need. Egress."

"I've been thinking a lot about our conversation last week. About what real freedom is. Yes, what we need now is egress. But, when we get out of here, will we truly be free? Will we be able to live every moment to its fullest, to speak truth to power, to love unconditionally? How can we ever live up to our new standards of freedom?"

"Exactly," Shane replies. "It's not going to be easy. It's not very free out in the free world."

"At least we can try, always try to live up to it," Josh says. "Even if we never get there, we'll never stop trying."

"I feel closer to freedom right now than I have for months, sitting here with you guys, just thinking about it."

The door opens and Ehsan comes out, carrying a box filled with dozens of small packages of butter, yogurt, cheese, cucumbers, and fresh tomatoes. I can see that it makes him happy to see us together and at ease. I ask him if he's heard anything from our interrogators, any news about us from the outside. He hasn't.

"I will call them," he says. "They should come soon. Now, enjoy your time together, I will be back in an hour." Another hour! It feels like he's giving us eternity.

"Will you ask them why they won't give Sarah a cellmate?" Josh adds.

"Yes," he answers, "they know it's wrong, but they still won't give me an answer . . ."

"Whose decision is it?" I ask. "The warden's?" We've been trying to get to the bottom of this question for months.

"No," he says, "not the warden's. Your interrogators."

44. SARAH

I have a purple headscarf that Leila delivered to my cell a few weeks ago along with some letters and books. It's the most colorful thing I've been allowed to have, embroidered with beautiful yellow, turquoise, and silver threads, with a fluffy fringe at each end. The colors lift up my imagination like a magic carpet. When I fly on its back, I can see the beautiful Iran that the guards love to boast about, the Zagros Mountains and the Caspian Sea, the Iran I wish I knew instead of this place.

"It's snowing," a guard named Maryam says, standing at the threshold of my cell door. "No *hava khori* today . . . You'll get sick."

"Maryam," I say in my broken, hard-earned Farsi, "I am not a child. I am a thirty-one-year-old woman. I want to go to *hava khori* and see Shane and Josh. Now!"

Suddenly, she reaches out and slaps my face. Without pausing to think, I reach out and slap her back.

"Man bache niistam!" I yell. *I'm not a child!* *"Man ensanam,"* I continue, pointing to my heart. *I'm a person.*

We stare at each other for a second. Neither of us slapped hard enough to hurt — just hard enough to make a point. Both of us are small women in our early thirties. Maryam has three children and an annoying propensity to treat me like one of them. The slap was the furthest I've ever pushed any of the guards, but she slapped me first, which we both know she's not supposed to do. Instead of slamming my door and walking away like she should, Maryam pauses. A glimmer of sympathy, even respect, softens her face. She tells me to get ready for *hava khori*.

Seconds later, tiny flakes are falling on our upturned faces. Shane, Josh, and I are standing hand in hand, feeling the soft, cool snow melt and trickle down our cheeks.

"Look at that one!" Josh says, pointing to a flake on my shoulder that immediately disappears. The three of us snuggle up under a wool blanket. I tell them about slapping Maryam, and Josh recounts the first time he slapped a guard's hand off his shoulder. Soon after that, the guard stopped pushing him down the hall. Shane recounts how he made a guard open the small window in their cell door by putting his foot in the door and staring him down.

"It reminds me of what Douglass wrote," Josh says, "about standing up to his slave master." A few weeks ago, we received the first stack of books we've been allowed from our families, including Frederick Douglass's autobiography, which the three of us have been passing back and forth ever since. There couldn't be a better role model than Douglass for standing up to power. After he stood up to his brutal master and fought back, Douglass was never beaten again, even though he remained a slave for several more years.

"The guards can sense it," Shane adds, "when we stop being afraid."

Over time, we've come to challenge and break every rule we can possibly get away with. A deep, long-standing contempt for illegitimate authority is one thing Josh, Shane, and I have in common. It's formed how we've responded to our captivity, collectively and as individuals, since the day we were captured. Now, after almost six months, we're faced with having to develop a strategy for long-term incarceration.

Everything we do, no matter how small, we try to do disruptively. I cover the peephole on my cell door with blue shampoo so the guards can't see me, stuff my extra clothes into the light fixtures for a few moments of darkness, sleep naked, refuse breakfast when they bring it and demand it later when I finally manage to drag myself out of bed. I shout, I laugh, I yell, I live as loudly as possible! This behavior leads to plenty of drama and petty confrontations with the guards, but it also keeps me feisty. Anything is better than going limp. I'm determined never to let prison break me.

The guards and interrogators counter this by creating an arbitrary, even irrational climate designed to sap our energy and make it impossible to develop an effective strategy of resistance. Some dole out tons of extra food while others refuse to give an extra minute or two in the shower. One month twenty-five letters might arrive from my mother — the next, only four. Any objections are answered with lies ("That's all your mother sent this month") and more false promises ("I promise to bring more next week if she sends any").

Despite our vigilance, at times we grow weary of these games. About a week ago, Nargess opened my door to hand me my lunch. Suddenly, another guard called her from down the hall and she left in a hurry. Standing with my food in my hand, I noticed a narrow, open crack in the door's seam, which usually let in no light. She'd left my door open. *What difference does it make*, I thought to myself, standing frozen with my eyes fixed on the crack, *if my stupid door is open?* In the hallway outside my cell, there is a video camera mounted on the wall. This hallway leads to another, wider hallway with another camera mounted on its wall. With no help from the outside, there's no possibility of escape.

Still, how can I close the door of my own cell? In *Discipline and Punish: The Birth of the Prison*, Michel Foucault wrote that modern prisons are modeled after the panopticon, an architectural design that stations the guards in a dark room in the center of the pod, with rows of cells arranged around them. That way, a prisoner never knows when or if she's being watched. Over time, she stops wondering. Once the omnipresent eye is internalized, the prisoner becomes her own jailer.

For months I'd had dreams of that damn door being left open — of magically walking out to freedom. Now that it actually was, I couldn't deal with it. *Fuck it*, I thought, *I'll just close it myself. No one will ever know. So I did.*

In this environment, sanity means balancing on a tightrope between acceptance and resistance. If you fight every battle, you'll exhaust yourself. If you stop fighting, you'll slowly lose your own self-respect. The important thing is that over the last few months I've managed to take back my own story. I'm no longer that madwoman sobbing and beating at the walls — I do things my own way. The one, real freedom I have is in choosing how to react.

The next morning I wake up early and sing "Gracias a La Vida" by Mercedes Sosa at the top of my lungs. When Nargess barges in and tells me to shut up, I sing even louder. The prisoners down the hall whistle and clap in approval. This place can't take away my gratitude for life. It can't stop me from being me.

45. SARAH

"How have you been, Sarah?" It's him, Father Guy.

"How do you think I've been?" I snap back.

I stand next to Shane and Josh in the interrogation room for three seconds of interminable silence. Being blindfolded has become second nature by now, so we stand with our heads down, not bothering to look at each other.

"How dare you?" I yell. "How dare you just waltz in here and casually ask me how I've been? How dare you ask me that?"

Shane grabs my hand and squeezes it gently. I have to stop talking because I feel like I'm choking. I can't get enough air in my lungs. I hate him. I fucking hate this man.

And I love Shane and Josh. I feel so proud standing with them as one unified, loving force. We outnumber him, I think, and we're clearly the ones standing on higher ground.

"Where have you been?" I ask angrily. "Why did you leave?"

Since Father Guy stopped coming here three months ago, my whole attitude toward prison has changed. I have vowed never to let anyone in here have as much power over me as he had.

"Sarah, they have put me mostly on another case, so I am not asked to come here anymore," Father Guy says. It makes me cringe to hear the self-pity in his voice. "I am sorry that you are not well," he adds.

His voice no longer sounds honey sweet and fatherly to me; all I can hear is the fake formality he uses to try to cover up his weakness and guilt. Whatever kind of sick friendship I had with this man has been over for a long time, and I won't stand here now and let him fuck with us.

"You should be ashamed," I spit at him. "Why do you pretend to care?"

He doesn't respond. His silence feels good. Being angry feels good.

I'm not looking for comfort anymore — I just want revenge. Even if it's only a fraction of the emotional pain he made me feel, I want to do everything I can to make him suffer.

Josh breaks the silence. "So, why are you here? What do you have for us? Any news?"

"Well, they've captured Abdolmalek Rigi, the leader of Jundullah, a terrorist organization, and he's confessed to taking money from the United States."

"Yes," Josh replies, "we saw that on TV. What does it mean for us?"

"Hopefully it means that they will let you go now that they have the real thing."

We all fall silent as that new, dangerously hopeful thought sinks in. I feel the seconds pounding in my brain. Soon, Father Guy will be gone.

"I have come here today because the judge has made a decision."

"The judge?" At the mention of that word, Shane, Josh, and I jerk our heads and lean forward like hungry dogs straining on short leashes. For a second it seems possible that his next words might set us free.

"The judge has decided to let you call your parents today."

"What? Why didn't you tell us before? Oh my God," Shane says.

"Oh thank God," I whisper as I bend my head toward my knees and try to focus on what's happening. All these months, I would have given up all my privileges, my bed, my TV, DVD, extra food and books, just for the chance to tell my mom that I love her. We're told that each of us will be allowed one call, five minutes in length, and that we are only allowed to talk about our health, treatment, and conditions. I repeat her phone number under my breath three times, relieved that I can still remember it. I had lost hope that this would ever happen.

The phone is bolted to the wall at the end of the corridor. Father Guy picks up the receiver, dials in some numbers off a phone card, and hands it to me. I want to sound brave, I think, listening to the phone ring. I'll tell her about our conditions matter-of-factly, but I won't let myself break down. I'll tell her how proud I am of her for fighting for me, what an incredible mother she has been . . . everything I've been dying to tell her for so long!

The ring is the most distant sound I've ever heard; I imagine it stretching out like blue vapor across two continents, thousands and thousands of miles. The sound of my mother's voice on the answering machine hits me like a wall. I take in a huge breath . . . waiting for the beep.

"Mom, it's me," the words run rapidly from my lips. "They're giving me the chance and I'm calling you from prison. I'm okay and I'm coping. I'm still in solitary confinement and that's really hard. I see Shane and Josh for an hour once a day, and they're together . . . I have books, good food, and some letters. Everything you're doing means so much to me. We know that we don't know the details, but we know a lot — it gets through to us. Mom, I never knew my own strength until now. I'm as strong as I have to be. You're an incredible person and I know you're doing everything you can. You don't have the power to make this decision, Mom, no one does. No one knows how we're going to get out of here. We're learning a lot, trying to make the best of this time. You're gonna be amazed at how good my math and vocabulary are. Tell everyone I love them, our family and all —" The phone cuts off.

"Argh!" I yell. "You have to let me call her again. I know she'll pick up this time. Please."

"Okay, Sarah, but you have already used two of your five minutes. Go ahead."

I pick up the receiver, dial her number, and listen to it ring and ring. The phone picks up and I hear my mom's voice, urgent and trembling. "Sarah?" I begin to cry.

46. SHANE

As I wait to go to the phone to call my mom, I realize that I can't remember her number. I strain my brain to pull it up, trying to ignore the interrogators' banter. Then it comes.

I stand at the phone, blindfolded, and dial with Father Guy standing over my shoulder. The phone rings.

"Yeeello." It's a man's voice. It must be my stepdad.

"Jim?"

"No, this is Al."

"Dad?"

"Shane!" It wasn't Mom's number that I remembered. It was Dad's. He is laughing with relief. "How are you?" It's so great to hear his voice. He always sounds so happy.

"I'm good, Dad. Josh and I are in a cell together now, but Sarah is still alone. We keep telling them they need to give Sarah a cellmate. Make sure you tell everyone that." They told me I have five minutes to talk. The pressure of this precious time seeping away is killing me. It's freezing my brain. Why wasn't I prepared for this? "What's going on out there, Dad? What can you tell me?"

"Well, we're doing everything we can. We keep pushin'."

"Do you think the U.S. will do a prisoner exchange?"

"No, they say they ain't gonna go for that. Our main hope right now is court. Your lawyer says if they take you to court, he is sure he can prove you are innocent and they'll have to let you guys go. So we're pushin' for that right now." Oh no. That can't be all he knows. They've told us in their letters they've been taking all these trips to DC. They *have to* know more. There has to be more than hope for court. Maybe he is harboring a secret that he doesn't want to divulge on the phone. No, that's not it. I can hear it in his voice. That is *really* his biggest hope. Shit. Maybe Dad doesn't know everything. I need to call Mom. They have to let me call Mom.

"I'm going to try to call your mom on the other phone right now," he says. "Maybe that way you can talk."

He is dialing her number. The phone is ringing. So much time is passing.

"Okay, here she is."

I talk. "Mom? Can you hear me?" I hear nothing. The seconds are rushing by!

"Dad, can you give me Mom's number? I'll try to call her before my time is up."

He does. After he says the first digit, I remember the rest.

"I love you, Dad. Don't worry about me, okay? I'm fine. Tell Nicole and Shannon that I love them and that I think about them all the time. I'll see you soon, Dad."

"We're okay too, Shane. Don't worry about us. Hang in there. I love you."

I hang up. "Can I call my mom?" I ask Father Guy.

"You only get one phone call," he says.

"But I didn't use up my five minutes!"

"I will ask my boss." He takes me back to the interrogation room where Sarah is sitting and brings Josh to the phone to make his call.

When I enter, Sarah is arguing gently with Josh's interrogator. He rarely speaks, this guy. Josh says he was stone cold through most of his interrogation. Now, he is talking. "We want to give you a cellmate, Sarah, but the prison won't allow it. It's against prison rules to put foreigners with Iranians. If there is another American in the future, we will put you with her. No problem."

"That's not true!" Sarah says. She sounds really angry. "You guys keep telling me that, but it's not true. Ehsan, the officer, says the prison has no problem with giving me a cellmate, but they can't do it until you approve it. Why are you forcing me to stay in solitary like this?"

"Sarah," he says calmly and deliberately, "it is against prison rules. And anyway, you are not in solitary confinement. You see Shane and Josh every day. And you have a TV."

"Will you just talk to Ehsan?" she implores. "He is right out there in the hallway."

"No, Sarah. Be quiet."

"Can I go to the bathroom?" I know what she is doing and I'm nervous about it. This doesn't feel like a good time to piss them off, while we are trying to call our families. He lets her go and I hear her out in the hallway talking to Ehsan.

"Will you come and talk to him?" I hear her plead.

The interrogator goes out and brings her back into the room. He sits her down and puts a piece of paper in front of her. "Write what you just did and why you did it," he says in an authoritative voice. "Make sure you write the name of the person you talked to out in the hallway." I can hear Josh down the hall, shouting into the phone as if he were yelling across the ocean. He sounds happy. I think he's speaking to his dad.

"Write why you broke the rules and spoke to someone when I or-dered you not to."

"Just say you were going to the bathroom," I say to her. We're sitting next to each other, but I can see only her hands from under the blindfold. "That's what you were doing. Say prison officials told you you could have a cellmate."

"Shane, go to your cell," he snaps. He's pissed.

"But I haven't called my mom yet," I say.

"You can't call your mom. You broke the rules. Go to your cell."

Father Guy, the good cop, walks in. He asks Josh's interrogator to let me make the call, but he refuses.

"Shane, go back to your cell," Father Guy says in a softer voice. "We were going to let you call your mother, but because of your behavior, my colleague says you can't." Sarah beseeches them to let me call. A guard comes and takes her away. Josh isn't here. They must have brought him back to our cell after his phone call. Now, it's just me and the interrogators.

"I'm not going until you let me call her," I insist. "Just give me two minutes. I just want her to hear my voice, to know I am okay."

Father Guy asks the other interrogator again in Farsi.

"He says no," Father Guy says. "You must go. Get up."

I refuse. Outwardly, it's an act of defiance, but I know that actually, my refusal is what they want. They have me. It's just me and them, and they know they will win what they are really after: my humiliation. For so long now, I've tried to maintain a tough exterior with them, to convince them they can't break me. But now this hard-ass has me in the palm of his hand. He has presented me with a choice and he has me checkmated. Either way, he wins. If I leave and go back to my cell, or even if they drag me away screaming, I will regret it. He will have the victory of refusing me the phone call and I will have nothing except my pride. And I will spend days or weeks punishing myself for letting my pride win over the chance to speak to my mom.

So I make my choice. I bow my head over the desk and I start to cry. "Please let me call her," I whimper. I've submitted. I've chosen humiliation.

"*Chi?*" Father Guy asks the other interrogator softly, again petitioning on my behalf.

The other interrogator says something to Father Guy, and he translates. "Okay, you can go. But only two minutes." I wipe my face, stand up, and walk toward the phone.

"You must not mess with my colleague," Father Guy whispers to me genially as we walk down the hall with his hand on my shoulder. "He is very serious." Despite his seeming friendliness, I despise Father Guy just as much as the others. He made me grovel just as much as that asshole. He just has a different approach. In fact, out of all of them, I think I loathe him the most. I hate him for acting like he has a conscience. I hate him for making Sarah believe he is good, when he just has a different strategy. I hate the way he says, "We are friends," or the way, months ago, he acted as though I had betrayed him by using an illegal pen—"I *trust*ed you." I hate the way he acted like he was hurt earlier when Sarah scolded him for not giving a shit about us. All he ever gives us is a soft voice—nothing more. He did show Sarah a few pictures once, but that was just to manipulate her. And it worked. Up till now, she has defended him.

"You guys have to trust me," she said to us once. "He can help us." This man who will never show his face will help us? I've tried to tell her that—"He is an in*ter*rogator in a po*lit*ical prison"—but she has always just looked at me like I didn't understand him. I had to watch him pull her in. I had to watch her *like* him. It made me sick and sad.

At least she got angry with him today. She's become so acquiescent with them, but today she felt like the person I know again, the one who doesn't take shit from anyone. Now he's dialing my mom's number on the phone.

"Hi, Mom, this is Shane." From the first moment, I know how inadequate this is going to feel. After not speaking to her for more than half a year, how do I greet my mom other than how I always greet her when I call?

"I love you, Shane." This is the first thing she says.

"Mom, I'm strong physically and mentally. I exercise a lot. We get extra food from the canteen." Does she believe it? What is she hearing in my voice? "They only gave me two minutes to talk."

"I want you to know that we are all doing okay," she says. The strength of her tone uplifts me. "I'm okay. Dad's okay. The girls are okay. And you have amazing friends, Shane."

"What's going on out there? Is our government doing anything?"

"They are, but there is only so much they will do." In her voice I hear her disappointment in not being able to give me something to be hopeful about. I've been saying all along that our government isn't going to get us out, but now I realize that maybe I was saying those things to keep my hopes down, because when she says this, I can hardly believe it. Deep down, I *did* believe our government would save us. She talks like they — our families — work alone, apart from the government. "We're coming up with new ideas all the time, Shane," she says. So, it's our families and friends versus the government of Iran. We're going to be in prison for a very long time.

"Your time is finished," Father Guy says, holding his finger over the receiver.

"Mom, I have to go. They are going to cut the call. Mom, I love you. I know you won't stop until we are free. I'm worried about Nicole and Shannon. Tell them I love them."

"Your sisters are okay. They are just worried about you. Shane, we are doing everything we can. This is going to end. I don't know when, but I know that you will be free someday. Know that we will not stop until that happens. I love you, Shane."

"I love you too, Mom. Bye—"

All I feel is loss. My parents have been taken from me all over again. There is no clear path to our freedom. There are no secret negotiations. We are here and no one knows how it will end. How can that be?

"There, I gave you *three* minutes," Father Guy says, as if seeking praise. I give him none.

47. JOSH

The first name I learned in this prison was Ehsan. The first Farsi sentence I learned was *Ehsan mikhaam. I want Ehsan.* Ehsan's dignified voice travels through the prison as if its walls were made of papier-mâché. His voice isn't that loud, but it's distinct, and I've grown acutely sensitive to it. At any hour, his voice shoots me onto my feet to petition to the nearest guard, *Ehsan mikhaam.*

In solitary, I'd count down the forty-eight hours until his next shift at 8 p.m. every other night. I like him. I genuinely like him. I tore off www.freethehikers.org from my mother's letters and insisted he look me up. Once, though most guards were wary of entering a cell, Ehsan sat on my bed. We chatted about his studies and his romance. Another time, he told me he had visited, but I'd been sleeping. "Ehsan," I remonstrated, "if you visit, wake me up." He witnessed my deepest desperation and helped me out. Ehsan is the only guard I almost consider a friend. Recently, he gave us pens and paper, items that before now we've never been allowed to have. He has even ordered the other guards to keep us supplied.

But Ehsan disappoints me when politics enter into our relationship. I've insisted he contact my family, but he does not. He won't even tell me news. He works for this dungeon, and that pisses me off. He should be held accountable for supporting this corrupt system. I wish he'd be ethical and quit, but, if he quits, he'd wouldn't be around and I'd miss him.

I'm in the shower located down the hall from my cell when I hear Ehsan's voice nearby. I've not seen or heard him since that phone call a few weeks ago. I dry off and ring the bell for a guard to come.

A guard opens the door. *"Ehsan mikhaam,"* I say. My cell is to the left, but Ehsan's voice comes from the right. The guard hesitates. So, without his approval, I stride confidently to the right. I inch my blindfold up and make eye contact with Ehsan seated in a chair. He's supervising hallway six, yelling at a prisoner to hurry up on his way to the bathroom. Ehsan straightens in his chair, surprised by my approach.

"Why don't you come to my cell anymore?" I say. "We need someone who speaks English. You're the only one. You know I miss you when you don't visit."

"Josh," he says, his face immediately slackening in empathy, "I cannot talk to you anymore. I am no longer the guard officer. I've been transferred. I'm only working here today as a substitute guard because of the holiday. I'm not allowed to talk to you. It is orders from the prison boss. Your interrogator grew angry when I told them that Sarah should not be in solitary confinement anymore. This is my punishment." He pauses and looks at me softly as if to say he's sorry. "Josh, do not tell anyone that I told you this. I can't talk to you." Then he turns away.

Tears almost come to my eyes. I lower my blindfold to hide my face. I've gotten used to so much, but this doesn't feel fair. Ehsan's leaving. I reach out to shake his hand, and we say goodbye.

48. SHANE

The three of us are sitting in the corner of a large conference room within the prison grounds. The usual pictures of Khomeini and Khamenei adorn the walls. There are little microphones at each seat at the large table. Dumb Guy has told us we will be meeting with an ambassador "from the region," but he won't tell us what country he is from. There is a news camera pointing at us from the Islamic Republic of Iran Broadcasting (IRIB) network. The three of us have already discussed how careful we must be in this meeting not to say a single line that can be taken out of context and used on television to sound like we are admitting to spying on Iran or entering illegally. Dumb Guy has suggested we apologize for crossing the border, but Josh and I insist none of us apologize unless we are shown evidence that we did, in fact, cross before we were summoned by the soldier.

I am more nervous about the video camera than I am excited about who we are about to meet. "Just be careful to stick to factual statements," I whisper to Sarah and Josh. "If we stick to what we know to be true, we will be fine."

In walks an entourage of officials. At the front is a man dressed in a flowing white robe and a coffee-colored, smartly tied headdress embroidered with shades of burgundy and brown. As soon as he enters the room, the air becomes redolent with sandalwood, an instantly calming scent that reminds me of the diwans of Yemen. He walks toward us, smiling warmly. His lightly bearded face is gentle but self-possessed and he gives off an aura of power — not the same kind of power his Iranian escorts display, but one that is regal.

He shakes our hands. "My name is Salem al-Ismaily, and I am here to represent His Majesty Sultan Qaboos of Oman," he says. "He would like to send you his warmest wishes."

We all sit and the three of us are expectant. This man obviously has a mission. He speaks to us in English that is only slightly accented, further extolling the graciousness of His Majesty the Sultan as well as the generosity of Supreme Leader Ayatollah Sayyid Ali Khamenei and President Mahmoud Ahmadinejad. This goes on for far longer than is comfortable. I'm worried this meeting will be over soon, so I am anxious for this guy to get to the punch line.

"So, why did you come here?" I ask, cutting into his preamble.

"I came to see how you were doing and to connect with you as human beings," he says. *That's it?*

His soliloquies ramble from one topic to the next, none of them having anything to do with us or our detainment. The talking points he runs through are awkwardly out of place. It's as though he is hitting Play on a tape he runs through whenever he meets with Americans. He keeps telling us that people in the Middle East aren't a bunch of terrorists, which I find condescending — he does know I speak his language and have lived in the region for years. He condemns the pitfalls of the Oslo Accords. He points out, as people in the region often do, that "Sarah" is an Islamic name. "In fact, did you know that Jews should be called Sarites rather than Semites?" he asks us. Jews take their religion from their mother, he explains, not their father, and Sarah was the wife of Abraham. He bemoans the fact that the Middle East is ruled by regimes that don't listen to their own people, ironic given that he is representing a sultan. He explains to us that he has been doing this work for a long time, negotiating between countries, trying to bring about peaceable solutions.

The work most dear to his heart, he says, is helping Americans understand the Middle East better. He gives a scholarship every year to Americans to study in Oman. "You know how I got started in this work?" he asks. "I used to live in the U.S. Then, after September eleventh, they start changing things at the airport so that anyone from a Muslim country had to be questioned and fingerprinted. It was humiliating and it made me so angry. I decided I would never go back to America unless it was a life-or-death situation.

"God must have been listening to me, because two weeks after September eleventh, my daughter got a brain tumor. It turned out that the *only* doctor that could help her was at UCLA. So I had to go back. At the airport, I found myself back in that line, waiting to go through the whole ordeal. A woman came forward and asked if I was Salem al-Ismaily. I thought she was going to take me aside for questioning, but instead, she took my daughter and me to the front of the line. I explained our situation, and a man overheard us and offered to drive us all the way to the hospital. The hospital even gave us halal food. I was amazed by such kindness. I then realized that I had been mistaken, getting trapped in my anger, and I decided to dedicate my life to helping these two cultures understand each other."

Sarah, Josh, and I keep trying to bring the conversation back to our case, explaining what happened when we were captured, but every time we do, his eyes glaze over.

"You know what?" he asks. "I have never looked at the details of your case, and I don't want to. I am not concerned with whether you are guilty or innocent." I'm starting to see a rhythm to his speech. Sometimes he speaks to us. Other times he speaks to the Iranians, through us. "What I am concerned with is bringing an end to this whole situation that works for *all* sides. This has to be a give-and-take."

"So what is going on?" I ask him. "What is the give-and-take?"

"I can't tell you anything about that," he says. "You just have to be patient. These things take time."

"Are there negotiations with the U.S.?" I press.

"Look, I don't want to get your hopes up. Let's just say I am trying to help you. Okay?" He smiles. We have learned nothing from this man and he doesn't seem all that interested in the particulars of our situation. Yet there is something comforting about him. Despite myself, I feel the little cavity of trust in my chest opening up.

One of the Iranians motions to Salem that time is up.

"Well," he says, "I have brought gifts for you from the sultan." His colleague hands us each a bag with stylized Arabic writing on it. Inside, there are ribboned metal boxes, bound in leather with an ancient map of Oman on the lid behind glass. They are filled with dates, neatly arranged and each stuffed with cashews, pistachios, and dried fruits.

"We know that here, like prisons everywhere, food is not allowed to come in from the outside," Salem says. "But we have made a special agreement with the government of Iran, and they have promised us that you can have these."

I pull another, smaller box out of the bag. Inside is a designer Cartier watch embossed with Roman numerals. Sarah's is pink.

After we do this, the IRIB reporter asks Salem for an interview. The Iranians aren't rushing us out. They obviously want us in the background, happy with our gifts and chatting with the other kind-faced Omani man who came with Salem. Josh apparently notices this and removes all of the gifts from the table.

The cameraman positions Salem about fifteen feet in front of us. Salem stands with his back to him, facing us, as the cameraman sets up his tripod. He just smiles at us, as if he is imparting a secret. "Don't worry," he says, looking at Sarah. "This is going to end." There is some kind of connection between them. Sarah is clearly hopeful about him. Why is she becoming so trusting of people in power? What happened to her old suspicion of men and authority?

"I know you will help us, sir," Sarah says to him. "You have God in your heart." She never used to talk about God like this. Prison is changing her. It's starting to worry me.

"Please don't," he says bashfully. "You are going to make me cry."

As he begins the interview, his colleague, sitting with us, starts to show Sarah and Josh some pictures of his daughter on his cell phone. I sit off to the side, staring gravely at the camera from behind Salem. The least we can do is appear unhappy for TV.

The reporter takes his position and says to Salem, "Now that they have admitted guilt, what do you think should happen with these Americans?"

Immediately, I flush hot and start shouting. "What did you just say? We have not admitted guilt! We never admitted anything! You lie on camera just like that?" I am shocked by the ease at which truth is flung out the window. All of my rage at the simple farce of our situation pours out. My heart is pounding. Everyone in the room is stunned. The Iranians look dumbfounded as I continue shouting, and no one attempts to stop me or the interview. Sarah has a look of confusion on her face. "What happened? What is going on?" she says to me.

"The reporter lied!" I shout, facing her. "He said we admitted guilt!" We need to make sure the interview is useless. They will never air it with us shouting in the background. Sarah starts yelling. Josh starts shaking his head and waving his hands, aware that our voices probably won't matter as much as our images, since IRIB rarely ever airs the audio of their interviews. They usually just show them on screen, dubbed over with the newscaster's own version. Despite our commotion, the interview continues. As we yell, Salem calmly says to the camera that this case needs to be decided in the courts and that it should remain a judicial issue, not a political one.

When it ends, I struggle to show my gratitude to Salem. Our good-bye is awkward. As Dumb Guy walks us through the prison grounds, back to Section 209, I wonder if I just ruined an opportunity for freedom. What if I scared him off?

"I liked him," I say somewhat dejectedly to Sarah and Josh.

Dumb Guy takes the gift bags from each of us before we put our blindfolds back on and enter Section 209. "I thought we could keep the dates," Josh protests. He loves dates.

"We will bring them to you," Dumb Guy replies. He never does.

49. JOSH

I've noticed that Shane has been particularly angry lately. When he wakes up groggy, I ask what's wrong, and all he says is "Prison." Did a particular guard piss you off? Was there something I did? Something Sarah said? No.

I agree that prison sucks, but daily life now determines my mood. We've created a lot of activities and many of them we find fun—especially with our pens and paper. Now, I keep a correspondence with Sarah in Spanish; I create math lessons for her; Shane and I exercise for a few hours every day; we all take turns creating stories with fifteen new vocabulary words to present at *hava khori*; we study and test one another on Greek mythology and Islamic history; I'm memorizing William Wordsworth's poem "Character of the Happy Warrior" ("As more exposed to suffering and distress; / Thence, also, more alive to tenderness . . .") and am now enjoying *Lord Jim* by Joseph Conrad. Sometimes, by the end of the day, I even wish there were more hours to finish all my tasks.

What I look forward to most is *hava khori*. It's the time of day when I stop reading, writing, exercising, and being "productive" and just connect with my two friends. Increasingly, though, it doesn't work out like that. I find that more and more I'm mediating conflict between Shane and Sarah.

Sometimes, the smallest change in one of their facial expressions or tone of voice can deeply upset the other. Other times, they'll get frustrated because they expect the other to heal the pain of incarceration. Then they look to me when they think the other is being unreasonable.

When Shane and Sarah polarize, I sit in the middle, encouraging them to actually listen to each other. The conflicts can be about anything. Shane wants to share meals at *hava khori;* Sarah doesn't want food to distract us during these precious times together. Sarah and I want us three to read the same books at the same time; Shane doesn't. Sarah wants us to appeal for clemency and apologize, but Shane and I don't want anything that hints at admitting guilt. Sarah wants to live in a house in Oakland; Shane would prefer living in the nearby countryside. The content of the arguments vary, but the dynamic repeats.

In private, they both tell me what they think is wrong with the other person. Though they try hard to accept the other's "flaws," again and again, the bickering loops: "She's not acknowledging my pain." "He can't put himself aside for a moment, even while I'm alone in solitary confinement." "Why can't you just listen to me without letting your issues get in the way?" My father always told me that the first year after making a lifelong commitment is the toughest, and I'm sure incarceration doesn't help.

It reminds me of trying to solve conflicts between my parents as a boy. Whenever I bring Shane and Sarah together and they let down their defenses, I feel triumphant. When my efforts fail and a guard comes to take us back to our cell in the midst of an argument, I feel personally defeated. I know Sarah appreciates my contribution to our triad, and Shane has told me so much in our cell, but sometimes when I sit next to them as they argue, I'm amazed at how stuck they are in their own little dyad — it's as if they don't see me, another person, sitting beside them. They take my support for granted and are lost in themselves.

I wrack my brain for an escape from our routine. I propose that we celebrate as many holidays as we can think of—as creatively as we possibly can, even disregarding the actual calendar if we feel like it. The night before Palm Sunday, each of us creates a personal system for palm reading. On Sunday, we sit extra close and have a half-hour of levity reading one another's palms under the open sky. We laugh at our glorious futures and the imminent triumphs fated for each of us.

A few days later we decide it's Ash Wednesday. We smear our faces with the chalky, ashen *turbah*. Then we sit in our triangle and enjoy one another's company, frequently laughing at how ridiculous we look.

On Friday, we celebrate Passover—a story of liberation and justice. I arrange a makeshift Seder plate: a hard-boiled egg, salt water, a fish bone as a lamb shank, lettuce as a bitter herb, and dried, flat Persian bread for matzo. Halfway through the Passover story, unaccountably, Sarah gets the giggles and can't stop. She curls into Shane's arms, uncontrollably laughing and embarrassed by her girlishness. Shane looks at me, clearly uncomfortable and frustrated by Sarah. I call off the meal, a little ticked. I put effort into making the Seder special, and I refreshed my memory by reading the story of Moses in the Quran. Passover is my favorite Jewish holiday.

I peer down at my pathetic Seder plate, then over at Shane and Sarah balled together, and then I look over the walls to the cloudy spring sky. I quickly let it go. It was a nice week, but these distractions couldn't last forever. I realize that our holiday season is over. I had hoped for an Easter egg hunt on Sunday and was brainstorming how to celebrate May Day. But we could barely live in our bubble of distraction a whole week. Gravity pulls us down. "Prison"—as Shane exclaims grumpily in the mornings—tears us apart.

50. SARAH

Out of the corner of my eye I see a flash of a pink pant leg through the slot at the bottom of my door. A prisoner is on her way to the shower. She's only been in our hallway a few days, but it wasn't hard to come up with a name for her. She's the only prisoner I've ever seen wearing a bright pink jumpsuit, so I call her Pink Lady.

I'm sitting on the floor with my supplies for prison pie spread out around me. I have a plastic bag filled with digestive biscuits, which I am methodically crushing with my metal spoon. I add five or six squares of butter to the bag of crumbs and squeeze it in my hands until it forms a thick paste. Then, I shape the paste into a crust on my metal dinner plate.

Next, I place chunks of chocolate on the rim of my heater to melt. I use a spoon to mix the melted chocolate with dates, sugar, and butter into a thick, brown paste that I then spread on top of the crust. The first time I made prison pie for Shane and Josh, the three of us sat in a circle shoving huge spoonfuls into our mouths, almost unable to process how good it was. "Sarah," Shane finally said, "you could sell this. I mean, it's not just us — anyone would think this was delicious!" The last step requires help from the guards. I ring the bell and Maryam agrees to bring me a plastic knife. She stands and watches me while I cut the apple into fine slices, arranging them like a fan on the pie's surface.

A few minutes later, I hear Pink Lady's shower turn off. A guard walks past my cell to let her out, walking a few steps in front of her as they pass my cell. At the last minute, I decide to crouch down and peer through the slot in my cell door. Her appearance is even more surprising than the color of her clothes. She's a tall woman in her late thirties with bleach blond hair and tattooed eyebrows (which must be popular in Iran; even some of the guards have them). From my position at her feet, she looks statuesque, powerful, almost regal. I quickly make a noise — "Psssst" — and her eyes dart down under her blindfold to the slot I'm peering through. For a split second our eyes meet, she slows her gait, and her lip twitches almost indiscernibly. It's a smile.

The next day I'm in the middle of my exercise routine, doing jumping jacks and pushups, when I see a flurry of motion outside the slot on my door. I look down to find a tight ball of tissue paper on my floor. I grab it and immediately sit down with my back to the door. If the guards catch whoever threw this on the video camera, my door will burst open any second and I will eat the note before they can take it. But the guards don't come. Carefully smoothing out the crumpled note, I read:

> *Dearest Sarah, I am Zahra. Do you remember me? I have been very worried about you, my dear Sarah. They took me to a different section for talking with you, but now they have brought me back. Are you okay? Do you need help? I will talk to you tonight when the guards are sleeping.*

I can't believe it's her! Pink Lady is Zahra, the prisoner that Leila caught me talking to through the vent a few days after the prison was flooded with Ashura protesters. We've been in prison what, nine months now? That was almost four months ago that they moved Zahra. I've often wondered what happened to her, but I never expected to see her again.

Later that night, I wake to the sound of loud knocking on my wall. I hear my name being whispered in the hallway and crawl toward my cell door.

"Hello," I whisper timidly through the slot.

"Sarah," the voice replies, "I am Zahra. Do you remember me? I've missed you. Are you okay? Are you still alone? I've been very worried about you."

That night we devise a method of communicating with each other through notes written on scraps of cardboard. She will write with a pen she stole from her interrogators and I will use a small piece of metal I've fashioned from a tube of Vaseline that leaves a mark like a pencil. Josh, Shane, and I were actually allowed pens and paper for a short, blissful period a few months ago — but they were taken away after I was caught passing a note to my neighbor, a young woman named Hengame, who always sang to me and seemed to know a lot about our case. Shortly after I was caught, the guards raided my cell, taking my extra DVDs, all the study aids I'd painstakingly devised, and even some of my books and extra clothes. I knew we'd never be allowed pens again.

Zahra and I decide to hide our notes in the trash can in the bathroom at the end of the hall. She balls hers up in toilet paper and I stuff mine inside soiled-looking maxi-pads — places the guards will never look. When one of us has a new note waiting, we will let the other one know by three hard knocks on our common wall.

In prison you develop revolutionary patience—you wait for something that you know may or may not happen—with unshakable resolve. I sometimes wait for days or weeks for the right opportunity to pass a note or exchange a few words with Zahra. I wait till the right guard is working—the one who never bothers to check on me through my peephole—so I won't get caught writing. I memorize the guards' footsteps and the patterns they walk in through the halls. I save a small portion of beef stew in which I carefully soak a maxi-pad overnight, then let it dry for a day or two until it authentically looks like menstrual blood. If anything feels off, even the smallest detail, I abort the project. I know what I'm doing is risky, but I'm determined to outsmart them. I'll never let them catch me again.

These are the best days I've had in prison. For months the silence in my corridor has been broken only by the sounds of prisoners weeping. Zahra and her cellmates laugh and sing, choosing American songs like Michael Jackson's "You Are Not Alone" for my benefit. I sing back to them, feeling joy and connection. Zahra's also bold with the guards, sometimes making jokes, sometimes yelling at them. "I will not cry for these bastards," she writes me. "I will not show them my tears."

"We have to stop," I write to her one morning. "I'm afraid we'll get caught and they'll move you. Zahra, when we are both free, I'll come to see you in the Netherlands. We'll spend days together dancing and talking. We will be friends forever."

A few days later Zahra passes by my cell in a flurry and leaves another present balled up on my carpet.

They are moving me again—don't cry, Sarah! I don't know what will happen to us, but remember you are never alone here. Sarah, please remember that Iranians are not bad people. We love the American people. I love you. No matter where they take me now, I will try to find you. Remember to listen for me—I will call out your name at night.

I know Zahra's still nearby because, from time to time, I see her clothes hanging on the prison clothesline when I go out to hang up my own uniform to dry. My face lights up when I see her lovely pink jumpsuit. I run my hands across the pretty color and I sneak a few nuts or a piece of candy into her back pocket. It may not be much, but it's the closest I can get to her, and I want her to know I'm still here.

51. SHANE AND SARAH

Shane

Dumb Guy is at our door. He hands us a bag of brand-new jeans, shirts, socks, and sneakers with shoelaces. "Put these on, quickly," he says, and leaves. We've been planning for this moment. Our moms have told us in letters they were trying to come and visit us.

They recently put Josh and me in a new cell. This one is smaller, but it has a little three-by-five-foot bathroom in it. After Dumb Guy is gone, I go to the bathroom, where a tiny note, covered in microscopic writing, is wrapped tightly in plastic and stuck inside the outer lip of the sink. The little bundle is small enough to fit discreetly between my middle and ring finger and can be passed off in a handshake. Josh and I have practiced this many times. I tape the note to my penis.

They took Sarah's pen before ours, so we knew to hide one of ours before they raided our cell. I scribed this letter with our secret pen over several days while Josh stood watch for guards. It describes in detail what happened when we were captured and lays out a schedule of all of the events of our detainment, including prison transfers, hunger strikes, the arrival of books, and the changing schedule of meetings between Sarah, Josh, and me. About four months ago, the interrogators started giving us books sent by our families, so in the note, we make clear what we want to read and hope to receive. It describes our daily routine and lists our e-mail passwords in the hopes that someone will change them to prevent the interrogators from reading about our personal lives. It has a list of songs we sing that we want our friends and family to listen to.

We are transported in a van with fogged windows to a hotel in another part of the city. Large, unsmiling men with radios and bellhop uniforms take us to the fifteenth floor. I ask to use the restroom, where I untape the note from myself and put it in the coin pocket of my jeans. I scan the room for anything I can take — an impulse that has now become second nature — and find a comb in the toiletry pouch. On it is printed "Esteghlal Hotel." I pocket it and exit the restroom.

They line us up in front of a set of double doors, where we stand on red carpet, the kind movie stars walk down. No one has told us what is on the other side.

As soon as they open, I see our mothers standing before us like ghosts, shrouded in black and bearing flowers. Their faces show an admixture of sorrow and gripping relief. Like the female guards at the prison, only their faces and hands are exposed.

Lights and cameras are blazing everywhere. The next thing I know, my mom is in my arms. Suddenly, it feels like everything is okay. The smiles on Josh's and Sarah's faces disarm me even further. Our moms are suddenly real to us and we to them. I feel a warmth so pure that it awakens an old, lost part of me. I become loose in a way I haven't been since we were hiking up that mountain ten months ago. I pull the note out of my pocket and squeeze it into my mom's hand. As I hug her, I tell her to hide it. She tucks it into her bra. Then, she whispers into my ear, "I love you, Shane. I can't wait to have you home." In this moment, I feel halfway there.

Before we get to talk, men in suits sit us all down on a couch in front of a wall of cameras. Livia Leu Agosti, the Swiss ambassador, sits with us. Suddenly, we are in front of the world, a blazing spectacle. I don't understand how this is happening. After all those months of not letting us have pens in order to prevent any chance of communication to the outside world, why are they putting us in front of cameras, giving us a chance to say anything we want?

Our mothers stare into the cameras and thank the Iranian government for the "humanitarian gesture" of letting them come see us. Then Sarah, Josh, and I take turns answering the reporters' questions. What is the food like? Do you have any indication that you might be released? What do you want your government to do for you? Have you learned Farsi? How are you treated?

"We have a decent relationship with the guards," I say. "It's been civil." As soon as those words come out of my mouth, I regret it. People are tortured there. The guards beat people. Sarah has been in solitary for almost a year.

"What happened at the border?" a reporter asks.

"We never walked into Iran," I say, then stop myself. I know our interrogators wouldn't want us to answer that question. I feel an overwhelming need to self-censor. "We can't really talk about that," I say.

Suddenly, I realize why they put us in front of all these cameras — they aren't afraid of us saying anything they don't want us to. They control us. They didn't tell us what to say, but I have been afraid that if I say anything even slightly offensive, any plans they might have to release us could be canceled. They exercise the same power over our mothers. Their reach is long and deep. I stifle my anger at them and at myself.

When the interview ends, the reporters file out. My mom clings to me and smiles in a way she only could on seeing that her child was safe. It feels so good, but it also hurts. I don't know how much time we will have — one hour? — and my joy is mixed with a preemptive sense of loss. Too soon, we will be torn apart once again. This visit is a cruel gift.

Something is different about my mom. When I was growing up, she was a disciplinarian, a don't-make-my-mistakes kind of parent. She has always been a tough, no-nonsense kind of woman. Now she won't let go of me. Worry is deep-set in her eyes. At the same time, there is an out-of-place happiness there. It is the kind of sudden joy that could exist only for someone who has suffered enormously, boundless yet backed by a pain that keeps her eyes teary and quivering. "I hope they let you come home with us," she says. Her words make my heart drop into my stomach. She really thinks this is the end. I can see it in her eyes.

"I don't think that's going to happen, Mom," I say gently, squeezing her hand. Ever since we heard they might come visit, I have never even considered that possibility. The three of us believe we'll get out eventually, but we know the Iranians arranged this visit in order to hold us longer. By allowing us to see our mothers, they can claim they are making "humanitarian gestures." They can temporarily subdue international pressure and put the focus back on the United States to reciprocate.

The suits tell our moms they can remove their hijabs and be comfortable. They leave the room. At last, we are alone.

When Sarah takes off her hijab, I am stunned by how beautiful she looks. I haven't seen her bare head since I snuck into her cell that first month. Her hair is so long. I hold her hand and don't want to let go. Already, this has been the most time we've spent together since our capture.

Livia jumps up and walks around the room, lifting up garbage cans and carpets to check for recording bugs. "This hotel is very famous," she whispers. "This is where they bring prisoners to do videotaped confessions."

We ask her about the nuclear negotiations. We figure that our freedom is impossible if the nuclear talks aren't going well. Every time Hillary Clinton snubs Iran for its nuclear program, I mentally settle into the idea of at least another two months in prison. We hear about the nuclear issue every day on TV, but it is impossible to know what is going on by watching state television. Livia says she doesn't believe that our detainment is directly tied to the nuclear issue, but "it doesn't help." She is clearly frustrated at the way the U.S. government is dealing with Iran. "You can't just make demands to Iran," she says. "They will never respond to your government demanding they release you. It only makes things worse. They need to *talk* to the Iranians."

Changing the subject, she gives us two bags full of books, letters, and postcards from all over the world. Throughout all of our time together, a part of my brain never stops preparing for our return to the cell, thinking about what we can do to make our lives a little better once this whirlwind has passed. I know the authorities won't let us keep the letters, so I stuff as many of them down my pants as I can fit. My mom digs through her purse and gives me whatever I want. I take a pen, some herbal tea satchels, and a small Farsi-English phrase book. I go to the bathroom and arrange everything so it is evenly covering my thighs. My jeans are so stuffed, it is hard to walk.

Mom tells me about my sisters, and though she tries to hide it, I can see they are in bad shape. My imprisonment is chiseling away at them. She talks about my friends as if they were hers. Shon, who was with us in Kurdistan, is moving in with her. Dad is raising money through hog roasts in rural Minnesota and by raffling off Bobcat skid loaders. She and Sarah's mom have been living together for many months. My mom, who normally gets up and has coffee on the deck looking over her dogs and horses, now drinks it in front of a computer. She, who was angry when they paved the road in front of her house in the woods because it would bring traffic, has flown to New York and DC at least six times each. Now, she rarely goes hiking because she can't get cell reception. She closed her business as a canine massage therapist and now works on our campaign full-time. It's what she wants, she says. Don't feel guilty. I know that she will never stop and I know that she is the best person to do this—I have never known anyone as strong as her. As the minutes roll by, I answer every question about our conditions that has kept her awake at night.

My mom tells us that since they arrived, they've been kept in a nearby room. Yesterday, their minders took them on a "tour" of the market. There, a woman approached them and apologized for what her government has done to us as agents swirled around them. After the tour, they sat them down and played videos about how Iran's internal enemies were supported by the United States.

Josh is sitting with his mom looking out a window on the other side of the room. She wants him to get as much fresh air as he can. She looks nervous and excited at the same time and asks him about our lives in extreme detail. He answers every question calmly, and holds her hands in his. They alternate between hushed conversation and robust laughter.

Sarah and I, also in our own separate worlds with our moms, signal each other with our eyes and come together. We kneel down on the floor in front of them and hold their hands, forming a little four-person circle. They look expectant. I think that somehow they already know. "Shane and I are getting married," Sarah says, perking up and smiling sweetly. "Congratulations," they both say, somewhat nervously. I try hard to read their thoughts. Are they worried that we're being rash, planning out our futures in such an extreme situation? Are they, both divorcées, trying to hold back their own fears of marriage? Sarah shows them her ring of thread, and they soften, cooing about how romantic we are.

Then, after a pause, Sarah's mom, Nora, asks, "Can we tell the media?"

Sarah

I can't get over how strong my mom looks. The hysterical woman who visits my nightmares has nothing in common with the fierce, articulate person sitting in front of me. All these months I've been imagining her defeated and broken, but I was wrong. I don't have to protect her from what's happening to me. Even if I want to, I can't.

"I feel crazy sometimes, Mom. I cry for days on end. Then, I'm consumed by anger, I can't concentrate . . . can't control my thoughts. I need a cellmate."

I stare into my mom's warm, beautiful eyes. She reassures me that if we aren't released in the next few days, the families are really going to amp up the pressure. There's no one in the world that I can count on like this woman. I have no idea how she's found the strength to fight this battle, but she has. I see it in her eyes, in her set jaw and determined voice. No matter how hard it is — she's going to make sure I get out of here.

I tell my mom to follow me into the bathroom. We walk past the secret police hand in hand and lock the door behind us. "Mom," I whisper, "about four months ago I found a lump in my left breast. It's big and sore, and wasn't there before. They took me to the prison doctor but didn't allow me to ask her any questions. A few weeks later, they took me out of the prison to a real hospital in Tehran for a mammogram, but I haven't seen the results. They say I'm fine, but I'm not sure I believe them."

I lift up my shirt and guide my mom's hand to the spot that's been tormenting me. Her skilled fingers gently prod the area — after a few seconds she looks up at me and shakes her head.

"It's not cancer, sweetie. It's just a totally normal lump. You don't need to worry anymore."

My mom has been a nurse for over thirty years — I know she wouldn't lie to me about this. "Sarah," my mom whispers, looking intensely into my eyes, "I want you to keep complaining about this." We hear a knock at the door and Dumb Guy tells us to come out of the bathroom. "You're fine," she continues. "Know that — but don't stop demanding medical care. This is really important, okay?"

"Okay," I agree, opening the door. We walk back to where Shane and Josh are sitting on the couch, snuggled close to their moms. I marvel at what these women have been able to do for us, what they will continue to do. All three of them have been in the forefront of the media, advocating for us with unshakable resolve. Right now, we are all our mothers' children, seeking comfort and protection, but this will be over soon.

"Mom," I ask, "do people know who we are? Do they know our politics?"

I can see the doubt in her eyes. "Tell me what you want them to know, Sarah," she says. "I'll be your voice."

"I want them to know who we are, why we came to the Middle East . . ."

"Of course. Sarah, I don't think you can begin to understand how hard people are fighting for you guys. No one's going to give up, Sarah."

Shane

The suits bring us menus and encourage us to order. I ask for shrimp and a chicken sandwich and fries and Coke and coffee. I eat it all, as well as some of the fruit heaped up on the table in front of us. Josh sips a nonalcoholic beer. He and Sarah barely pick at their food. Every minute is starting to feel schizophrenic. Sarah sings songs she has written in prison. We don't know when this is going to end. It is starting to feel like someone telling us to have fun at gunpoint.

Two hours after we eat, the Iranians tell us it's time to go. We hug and kiss our moms and say every last thought we can think of as they usher us slowly down the hall. We wait in front of an elevator. When it opens, a man invites our moms to enter. They do, and we stare at each other, them on one side of the threshold and us on the other, a microcosm of the past ten months. Just as my heart starts to break, something inside me turns off. It's prison time again. I know that I never want to remember the look of defeat, loss, and deep fatigue on my mom's face. I'm going numb. As the doors close and tears well in their eyes, I know that this is the worst moment of their lives.

52. JOSH

On the way back from the hotel, I can feel prison approaching. Next to Sarah and Shane, I sit in the back of the van. I dig voraciously through the oversized zebra-striped bag full of letters and books and photos and clothes that my mother just handed me.

A childhood photo rests on my left thigh. In it Alex and I sit cross-legged under autumn foliage as toddlers; a giant football is on my lap. I wish Alex had come! He had written in a letter that he applied for a visa as the "male escort for the three moms." On my other thigh rests a photo of my friends smiling like sunflowers around a redwood tree. Jammed between my legs are letters from my friend Farah and my uncle Fred. I'm thinking of how my mother talked of Jenny—how she sends her love and buys my mother flowers. Since my first batch of letters last September, it's the only time I've heard anything from people besides my immediate family. As I sift through these piles, I realize I don't know what I'm looking for; I'm just bathing in love.

I've always loved my mother, but I've never before felt this warm and glowing with her. All our past squabbles and arguments, all my little judgments became irrelevant during our visit. All my fears about how she might be upset with me just disappeared. She's fought for me nonstop for almost ten months. Normally, we barely hug or kiss. It's just not how she raised me. But we didn't let go of each other for more than a few moments all day. It felt amazing to express my love to her. It's too bad it takes imprisonment for me to fully appreciate who she is.

The van turns sharply to the left, and all the letters and photos spew to the floor. The vehicle comes to a stop. I step out with my zebra-striped bag in hand to an unfamiliar scene. A dozen guards stand ready around a volleyball net in a parking lot next to the entrance of Section 209. Even more guards cheer from the sidelines. I recognize all of them. I've been wondering for months what type of ball game I hear from my cell.

The guards stop playing, put the volleyball down, and start chattering and pointing at us. Shane and Sarah are a few feet ahead, approaching the building. The guards are pointing *at me*. One guard jogs over to me.

For a moment I think he is going to invite me to play. This guard likes me. Many of them do. Sometimes, in the hallways, I play-wrestle with a guard we named Miscellany. Other times, I hold hands with a guard named Jon when we walk down the halls. No guards normally hold prisoners' hands, but I know male physical contact is normal in the Middle East. The janitor and I exchange smiles and a *"Salaam"* every time we see each other. I joke with a guard we named Peugeot about his favorite TV series, *Prison Break.* Sometimes this rapport allows me leeway, but the guard who runs to me from their volleyball court wants me blindfolded and inside the prison's walls. He pushes me toward my friends and the prison building.

When the other guards see what is happening, the rest of the volleyball court erupts in commotion — everyone arguing. I have no idea what they are talking about, but the debate is heated. Suddenly AK, the Ass Kicker guard, runs toward me. He carries the ball in his hands. He grasps my forearm, pulls me onto their makeshift court, and hands me the volleyball.

I gain immediate satisfaction from his actions. I maintain the conceit that I am slowly winning over the stoic and cruel AK with my charms. When he recently gave me a shoulder massage as I walked down the hall to *hava khori,* I felt as though I'd started making progress. I like to think that by being friendly and respectful with AK, I'm lowering his defenses, and he'll eventually treat me like a human being. But I never imagined playing volleyball with him.

I love seeing the guards divided over how to deal with me. It is so characteristic of this upside-down place that a guard who often jokes with me acts tough in front of his peers and pushes me toward the prison. Meanwhile AK, who earns his reputation by beating prisoners, invites me to play. I feel oddly honored to be singled out. I can hardly believe I was with my mother a half-hour ago, and now I'm playing volleyball with the guards. I brush these thoughts aside, grab the ball, and stride to the server's line as confidently as I can.

I bounce the ball a couple times and look around at my teammates. They are ready. I take a long look at my opponents on the other side of the net. Their knees are bent, their hands raised. *They* are waiting on *me.* The ball is in my hands. A package of ballpoint pens that my mother gave me rests precariously on my crotch. Clouds drift above us. The snowcapped mountains loom majestically beyond the prison, and I savor this moment of power.

My serve descends deep on the left side. They bump the ball out of bounds — our point. If we were playing by the rules, I'd keep serving. But I am content walking off the court to sounds of approval.

Inside the prison, they immediately take Sarah upstairs and Shane and me to a changing room to dress into our uniforms. Before we enter, a guard empties our pockets and waistline full of letters, pictures, and talismans that our mothers gave us. I want to carry the feeling of being in my mother's arms with me into my cell. I want to cherish all that she gave me and to remember the warmth that exists in the world. But no. They take everything except the four pens balancing on my crotch.

53. JOSH AND SHANE

Josh

It's now June, one month since our mothers were allowed to visit. Since then, Sarah's been miserable at *hava khori*. She has fallen into a deep depression. Nothing Shane or I do lifts her spirit. The morale of our triad is sinking steadily. To add fuel to the fire, the United States initiated more UN sanctions on Iran, and our hopes for freedom are crashing.

Since including me in his volleyball game, AK has been pissing me off. Once, on a whim, he made me leave *hava khori* earlier than Shane and Sarah. Then, the last time I saw him, he locked the window in our cell door for no apparent reason.

We are walking back from *hava khori* when Shane grabs an extra dinner tray from the meal cart, which we routinely do after they distribute meals. Immediately, I hear AK barking from down the hallway. He strides toward me, until he's just inches from my face.

In awkward Farsi, I try to tell him I'm angry with him for slamming my window a few days ago. I want an apology. But he doesn't acknowledge it. So I say it again, louder.

He inches even closer to my face, breathing on my forehead. I stand, wearing my blindfold and still waiting for him to apologize.

A female guard whisks Sarah away. I hear her ask in a concerned voice, "Is everything okay?"

"It's fine, Sarah, don't worry."

I turn back to AK. I'd been thinking about him closing the window for the past few days and I'm excited to finally be confronting him.

Instead of apologizing, he shoves me. I stagger back in my sandals and steady myself. He doesn't give me a moment to gather myself; he shoves me harder. Again, as I straighten up, he's waiting for me.

I cling to the nearest door. A guard we've named Tall Racing Stripe watches gleefully nearby. Something about the scuffle amuses him. AK pushes me again and I find a railing. Wherever he's taking me, I don't want to go. He looms above me, pushing me down the steps. In the stairwell, one of the nurses stands aside, wearing his usual, self-satisfied smirk as I stagger past him.

Shane

Josh is gone. I struggle to run after him, but I can hardly move. My shoulder is against the wall, and there is a guard in front of me and another behind me. I push and push. I lift my blindfold and keep struggling as hard as I can, pushing and pulling in sharp, frenzied jerks. Sarah is gone. Every instance of AK beating people is swirling in my mind.

I realize that one of the two guards trying to restrain me is the tall one we call Paper. I don't know the other guy. Paper has a kind of compassionate look in his eye. He's not angry. He is way stronger than I am and he is trying to calm me down. His softness almost sinks in, but I force it away and keep struggling. They pin me into a corner. There is no way out.

Suddenly, AK comes back up the stairs, reaches through the two guards, and grabs me with one hand. He tosses me to the floor like a kitten. Next thing I know, I am in a headlock. My whole body feels like it is shouting — in my mind I am shouting — but I'm not sure whether I am making noise or not. Suddenly, he is marching me down the hall with my head pressed against his side.

My feet are trying to keep up with him, but I keep slipping, which makes me fall into the iron noose of his arm and choke. His bicep must be as big as my thigh. As he drags me, his breathing is deep and labored. Halfway down the main hall, he stops, reaches under the guards' desk, and pulls out a billy club.

He drags me down the rest of the main hall, his stride long and heavy, then turns down our hall and flings me into our cell. In one motion, he steps into the middle of the cell and smacks the billy club against our little fridge with all his might. In the next instant it is above his head, ready to come down on me.

I cower instinctively. For the first time since this started, everything pauses. Sweat is beading on his forehead. His eyes are bloodshot and wild. He could break bones. I am terrified. I put my hands up in submission and he storms out of the cell.

My body falls down onto the bed in a huff of physical relief. I put my face in my hands and I exhale. Then I remember Josh. I know what I must do.

I leap up to the door and shout. "Josh *kujaast?!* Josh *kujaast?!*"

AK storms back to the cell like a bull chasing down a Spaniard. He flings open the door, charges into my cell, and raises his fist.

I tap my finger against my jaw.

"Hit me," I say in English. He winds his arm back threateningly.

"Hit me!" I shout. I feel crazed, but lucid. I know that I am taking away all of his power. If he can't frighten me, all he can do is hit me, and if he does that, he will be hurting himself. We are hostages, and hostages are currency, and currency is not to be damaged. Making him beat me is my only way to fight back. And it's my only way of keeping him off Josh.

He is getting worked up, but he is confused.

"Where is Josh?!" I say over and over again. I won't stop.

He raises his leg up as if to kick me in the chest. I slap my hands against my ribs like a man possessed. "Do it! Come on!" He has almost reached the tipping point.

Paper is standing at the doorway, watching. He looks unsure what to do, as though cognizant this shouldn't be happening but uncertain how to stop it.

"Where is Josh? Josh *kujaast?!*"

He pushes me and I fall to the floor. He is above me. It is going to start now. My body lets go, gives in to what I expect to come. I just tell myself, over and over again, *I know this.* Why is prison so much like high school, where power was stark and beatings happened regularly? *I know this.* I know it happens quickly and it hurts but not that badly. Unlike it was when I was a kid, this time I'm not going to lose because I'm not going to fight back. And I'm not going to fear him. I know that I have already won, and I think he does too, which only fuels his rage. I know that Josh will come back.

I stand back up. "Josh *kujaast?*" I say forcefully. Again he pushes me, harder this time. My foot catches on Josh's mattress on the floor. I fall backward and feel my head smack against the wall, bone on cement. He moves in toward me. I know this is it. This will be when everything becomes a black jumble.

Suddenly, I hear a voice. *"Kfaaya! Kfayaa!"* It's Paper. He has his hand on AK's shoulder. AK looks back at him and stands upright. They both walk out of the cell.

I touch the back of my head. A lump is starting to grow.

Josh

With AK back upstairs, one of the other violent guards grabs hold of me. The guard smiles at me like a child receiving a handful of candy. It's an ominous smile. This guard has been a jerk since the day I arrived. He grabs my arm and drags me to an administrator's office.

The administrator stands when I enter. The blood is still coursing through my arteries. The administrator sits me down in a chair. Then he sits down and asks, in Farsi, what happened. Using his office as my stage, I reenact the incident for him, pushing myself with my hands in a theatrical display of how AK pushed me. I throw in occasional Farsi words: "Guard! Problem!" I bang my fists together histrionically.

The administrator shakes his head in disgust, then takes a phone call. He doesn't understand that I was demanding an apology from AK for an incident a few days ago. He thinks the scuffle was for the tray of food Shane picked up in the hallway. When he hangs up, he orders his assistant to bring me dinner.

The administrator apologizes profusely. He assures me I can have as much food as I need. He tells me that AK is out of control and that he'll make sure we won't ever need to interact with him again. When his assistant hands me a dinner tray, he tops it off with two extra bananas from his desk. I can see his head spinning: How is he going to explain this to his boss? Who else will find out about this?

He puts a third banana on top of my tray. I can't wait to tell Shane about this. He ushers me into the hallway and directs me upstairs. Back in the cell, I see Shane. We hug, deeply relieved.

54. JOSH

One week later on our way to *hava khori*, I throw off my blindfold and see Shane and Tall Racing Stripe swinging punches wildly. Their feet are jockeying back and forth. Their right fists cock back. I kick off my sandals and race toward them. I'm unclear what I'll do — I've broken up more fights in my life than I've been in — but I run directly at them.

In this momentary boxing ring of a hallway, all are equal. No bars separate us. Repercussions feel irrelevant. Shane and Tall Racing Stripe are well matched. Neither one of them lands a blow. Tall Racing Stripe sees me running at him — it'll be two against one. The thought comes to me: *Now is our moment!*

I charge into the space between them. Tall Racing Stripe turns and swings at me. I dodge him and grab his wrists. We grapple. I hold on, restraining his anger. He writhes and frees his arms.

Meanwhile a hefty guard races down the hallway. He seizes Shane and locks him in a nearby, empty interrogation cell.

Tall Racing Stripe continues to swing at me though I'm not swinging at him. His punches are frenzied. I can't understand his rage. Though his fist comes inches from my face, I still can't believe he wants to hurt me. What have I done? His fist slams into the metal gate next to me. He pretends that it doesn't hurt and swings again. I need to get through to him. I've never heard this guard speak a word of English, but I shout: *"I'M not fighting you! What are you doing? I'm not fighting you!"*

A few other guards emerge from nearby hallways, and they calm Tall Racing Stripe down. I find my blindfold near the trash can and my sandals by the radiator. A guard lets Shane out of the room he's locked in and we walk down the hallway to *hava khori* in silence. It is 8:45 a.m.

This is a mess. Last week it was AK, and now this. After the incident with AK, Dumb Guy apologized and they started giving us *two* hours together instead of one at *hava khori*. The extra hour has been fantastic — and uplifting. We had asked Dumb Guy for more time at *hava khori* for months. After his concession, we concluded that defiance gets results. It's way more effective than pleading for sympathy.

But we need to pick our battles and be strategic. We can't fight every guard who acts like a jerk. My ideal — largely influenced by having read Gandhi years ago — is to be compassionate and defiant at the same time. Anger pervades everything in here, but we'll be miserable if we're always seething in anger.

I can't be mad at Shane for fighting with Tall Racing Stripe. Even Gandhi wrote that he would choose violent resistance over cowardice. That was what Shane did. Perhaps fighting, yet refusing to gang up on Tall Racing Stripe, was Gandhian. Moreover, even if Shane provoked everything, which I'm sure he didn't, I'd back him up.

I need Shane to know that I have his back like he had mine last week with AK. I also want him to know we must find another way to deal with asshole guards. My critique shouldn't feel like blame, and I'm worried he'll feel ganged up on if Sarah's around. So, before she arrives in *hava khori,* I say as gently as possible, "Shane, we're not going to fight our way out of here."

55. SARAH AND SHANE

Sarah comes in and sits down on the blanket.

"I just got into a fight," Shane tells her, and explains what happened.

"Baby, you hit him?" Sarah asks. There's accusation in her face.

"Of course I hit him. He swung on me."

"I thought you weren't going to fight back," Sarah says, her voice rising. "I mean, hitting a guard is serious."

"Sarah," Shane says incredulously, "he hit me. He attacked me. That is what is really serious. I never said I wasn't going to fight back."

"Shane, we all decided months ago that we'd discuss things that affect all of us. Look, I'm not saying we shouldn't stand up for ourselves, but you can't just start fighting the guards!"

"Think? Discuss? Did you want me to ask him, 'Please wait until I can consult with Sarah so we can make a decision about this'?! I didn't *ask* him to hit me!" Shane shouts. "There was no decision or planning here. A guard attacked, and I defended myself. You are supposed to be on my side — to stand by me against them no matter what. What the hell is this?"

Josh is sitting alongside them, watching silently with that defeated yet attentive look he sometimes gets when things between Sarah and Shane escalate.

"I am on your side," Sarah says. "I just want to know that you're on my side! Why is this different from when AK attacked you a few days ago? Why didn't you fight back then? I need to know that you'll control your impulses!"

"What impulses are you talking about?!" *She thinks I am a loose cannon,* Shane thinks. *She is blind, mired in her own suffering. Doesn't she see that with AK, we won? We put them on the defensive.*

"I'm talking about the impulse to get out your rage by fighting the guards. Shane, the situation we're in is bad enough. I can't stand any worse. What if they stop letting us see each other? What if they hurt you even more? Nothing is worth taking that risk. Nothing!"

"It's easy for you to sit here and say these things when you don't have to deal with this shit," Shane says. "Women guards don't beat prisoners. Male guards do. When I told you about this, you should have shown me that you stood by me, that you had my back no matter what. That should always come first."

"Shane, you know I never want you to get hurt! Look, the guards will be coming to take us back to our cells any minute. I need to know this won't happen again before we can talk about it. Can you at least promise me that?"

Shane's head is spinning. *Things have never been this bad between us,* he thinks. Just thirty minutes ago he felt strong. Now he feels weak and defeated. *I don't know if I can count on Sarah to be there for me anymore,* he thinks. *I feel trapped — outside and inside.*

"Okay, I won't hit back," Shane says blankly. *I just want this to end.* "I won't do anything."

But it's not okay, Sarah thinks. *A huge abyss has opened up between us. Is Shane right? Do I care more about myself than I care about my own partner, my future husband? Perhaps my fear is blinding me to his suffering. But what if Shane really is losing control? What if I can't trust him to make the right decisions for us? The one thing I'm certain of is I can't live without seeing Shane and Josh; I will lose my mind and may never recover. I have to put that before everything else right now. Even Shane.*

56. SARAH

A few weeks after my argument with Shane, I'm admiring a vibrant green grasshopper I just found languishing in *hava khori*. He's big and meaty, sitting in the palm of my hand like an elegant, armored tank. The door opens and Shane and Josh appear. They admire the grasshopper and tell me they have a special treat — a plastic cup full of sauerkraut they made themselves. They requested cabbage as one of our extra weekly vegetables, crushed up its leaves, added salt and water, waited a few weeks, and came out with sauerkraut.

"There was a ticker on the news," I say as we sit down, "about a British woman converting to Islam in Iran."

"Really? That's interesting," Shane says distractedly, laying out some food he's brought to eat with the sauerkraut.

I decide to be more explicit. "I'm thinking about converting myself," I say. "I mean, I believe in God, I have respect for Islam. If we all do it, it might help us get out."

"Baby, you know I can't do that," Shane says, looking up at me with worry in his eyes, "but if this is something you really feel . . ."

"I knew you would say that," I snap. Josh hasn't said anything. "Do you think I'm wrong? Josh? Tell me the truth."

"No, you're not wrong — it's just a big decision," Josh offers.

"And you have no idea how you'll feel about this later, when we get out," Shane adds.

"I — I don't really know if it's real or not. In many ways, Islam seems like the obvious path right now. My connection to God has helped me so much in here, but I don't want to do this for the wrong reason. If I do, God will know . . . I don't want to turn a beautiful thing into some kind of self-serving strategy."

"I doubt they would release you," Shane says, and then adds, "Well, maybe."

"What if I test the water, see what the interrogators say, would you two consider doing it with me?"

"I don't think I could," Josh says. "My family would freak out — I really don't even think I could fake it if I tried." Shane nods in agreement.

"Well, we need to try something. This is feeling like it could go on for years and I can't take it that long. I'm worried about my mom, about everyone. I'm worried about who I'm going to be when I finally get out of here. What if I'm so fucked-up, the rest of my life is destroyed? Converting, or something like it, could give Iran a way out, a reason to release us without looking weak."

There's silence as we eat our sauerkraut, bread, and beans. "We have to be ready to do something," I say, "especially if things get worse, something we may not be comfortable with." I'm building toward my next suggestion.

"What if I try to get pregnant?"

"What!?" Shane says. I've finally gotten their attention.

"Well, for sure they'd let me out of solitary."

"How would we have sex?"

"Out here, on a date night, under the camera where they can't see us."

"That's crazy, Sarah. We don't know what they'd do. Do you want to raise a baby in solitary?" Shane asks.

"No, Shane, I don't. I truly believe they would take me out of solitary or release me. Maybe I'm crazy, but I can't just sit here passively anymore. They are charging us with espionage! We all know what that could mean. We have to treat this situation like what it is."

Shane and Josh don't say much after that. They're simply not feeling as desperate as I am. *There must be a way,* I tell myself back in my cell. The little attention the lump in my breast got from the interrogators seems to have faded. Anyway, we have to think of a strategy that will get all of us out.

57. JOSH

Soon after being brought to Tehran, I figured we'd be released by Day 30. When thirty days passed, I still believed freedom was just around the corner — usually at the next holiday: the autumn equinox, Thanksgiving, then the winter solstice. Underneath these hopes lay my theory that if the Iranian government really wanted to show the firmness of its resolve, it would wait six months to release us. After half a year in prison, I added another two months to my theory.

When we reached the eight-month mark at the end of March, I finally reckoned with how indeterminate our incarceration would be. I promised myself I'd stop clinging to dates. *Freedom will come when it will come,* I now tell myself.

On July 13, more than eleven months after our capture, Iranian TV reported that the United States had released an Iranian nuclear scientist, Shahram Amiri. Iranian state media reports that he was kidnapped by the CIA in Saudi Arabia and interrogated in the United States. Then one day, he magically showed up at the Iranian Interests Section of the Pakistani embassy in Washington, DC, and flew back to Iran. For our purposes, the most notable thing about this story is that Iranian TV treats it as a national victory over the United States. Characterizing Amiri's release in this way might give the government cover to release us.

At *hava khori,* our usual dynamic plays out. If someone takes a hopeful stance, then someone else takes the opposite outlook. Today, Sarah expresses cautious hope, divining that "*something* will happen within two months." The more hopeful she is, the more skeptical Shane is. I'm wary of both of them jumping to conclusions too quickly. Eventually, I chime in, saying something about how we don't have enough information, that we can't trust the news, and that we're just guessing in the dark. *Freedom will come when it will come.*

It's now July 27, almost a year since we hiked in Kurdistan. Civilian life feels very far away. Oddly though, when I look back at all the uneventful weeks and months of languishing in the cell, the time just disappears. Though it felt like the longest year of my life, in retrospect this year of detainment seems to have flown by. I can only vaguely remember tidbits from the months of blankness. All that happened in March was that we called home. In all of April, we met the Swiss ambassador. In May, we met the Omani envoy and our mothers. In early June the interrogators delivered a load of books. In July, Amiri was released.

Today at *hava khori,* Sarah tells of another significant event.

"They let me make a phone call," she says.

"And?" I say, engrossed.

"You don't feel jealous? Do you?"

"No! What happened?" I say, relieved that she now has a privilege that I don't have.

"I spoke to my mom. They're planning big things for the one-year anniversary — vigils in thirty different cities. You had suggested to our moms they focus the campaign internationally, and they did. My mom sounds great." Her voice becomes more sober. "The interrogators told me that they wouldn't let you guys make phone calls, just me."

Why did they let Sarah call home? Everything they do is calculated, so there must be a reason. None of us speculate out loud what this could mean. In my mind, I can't help but feel optimistic.

58. SHANE

Food Guy is at our door, taking our order for the canteen. He comes every two to four weeks and unlike other prisoners, we don't pay for these orders because we aren't allowed to have money, which I assume is a precaution taken to prevent us from bribing guards. Because we aren't limited by money, we get whipped into a frenzy whenever he arrives. We struggle to get in as much as we can.

"We want apples and oranges and walnuts and four boxes of dates and chocolate and twenty packets of cookies and cabbage and ketchup and make sure you get enough for us *and* Sarah . . ." Shit! My stream of thoughts gets interrupted. I was saying all this in a halting stream of Farsi, but I start reaching the limit of my vocabulary, so I stumble and pause. He might leave. He's still jotting the words down at our door. He hasn't gotten down everything I've listed so far. Josh picks it up.

"And greens and, and peaches and milk and pomegranate juice and apricot juice —"

"And coffee," I cut in. "Nescafé. And walnuts, did we say walnuts? A half kilo of walnuts. And almonds." He starts taking steps back, still writing. "And cherries!" It's August. Everything is in season. "And pomegranates. And —"

"Besseh," he says abruptly, putting his hand up. *Enough.*

"And cake!" I shout. He steps back toward us.

"Cake?"

"Yes, cake. You know the little ones?" I am imagining the little cakes our old neighbor Hamid tossed into our cell so long ago. Sometimes I see big bags of those Twinkie-like cakes sitting outside some people's cells after the canteen order is delivered.

"Sarah's birthday is tomorrow," I say. Today is August 9. "We need cake." He is jotting it down.

"Like a birthday cake?" he says, holding his hands out in the shape of a full-sized cake.

"Yes," I say, not skipping a beat, "a birthday cake. We need a birthday cake." He nods, closes our door, and leaves.

Later in the day, he returns with a big white box. He opens it up, smiling. "Is this okay?" Inside, there is a big cake with coffee-colored frosting. There are thin pieces of carved chocolate and pink frosting flowers and glazed fruit on it. I thank him profusely.

Josh and I set the cake on the floor, marvel at it, and laugh. Sometimes, we just stare at it and say "Wow" breathlessly. We have a cake for Sarah!

We have to empty out our fridge to fit it inside. The next day I make little hors d'oeuvres. I smash walnuts and onions into the soft cheese we get for breakfast and I spread the mixture on digestive biscuits. I put a little plastic cup in the mouth of our thermos — the thermos Josh grabbed from the hallway months ago that they now fill with hot water for tea. The hot water warms the bottom of the cup and I drop a couple chocolate squares inside. I dip some almonds and cherries in the melted chocolate and set them out to dry. I love days like this, spent preparing little ways to make Sarah's day better.

We also sniff the bottles full of fermenting liquid that have been sitting under our bathroom sink. We discovered the possibilities of fermentation when, last winter, I went to take a date out of my box and it smelled vaguely like beer. We put the whole box of dates in a water bottle and by the end of the day, it was bubbling and frothing. But when we drank it two weeks later, it was vinegar, so we used it to make salad dressing by mixing it with yogurt. This time it smells right. We decided we let it go too long last time. Now it's still fizzy. It'll be like champagne.

We get to *hava khori* before Sarah. When she arrives, I tell her to keep her blindfold on.

"Happy birthday," I say, kissing her on the lips. "Today, we are going on an adventure. I have a whole day planned for you. First, we wake up in bed cuddling." I lie down next to her on a blanket I laid out ahead of time. "Now, we are going to go on a picnic, just you and I. I was going to suggest we ride our bikes, but it's such a beautiful day, so I thought we should walk. We are at Lake Merritt in Oakland." I guide her through one lap around *hava khori,* then sit her back down on the same blanket. Josh is sitting on the opposite side of the courtyard, quietly setting up the next stage. I feed Sarah the snacks I made, one by one. She marvels at each one and tries to guess its ingredients. I rub her belly and stroke her hair as if we were two lovers dallying in a park.

"So, I thought we'd just go back to your mom's apartment now and have a nice evening, just the two of us. She's not there. Josh will probably stop by later on."

"Okay, that sounds nice," she says in a self-aware play-acting voice, smiling. I walk her around the perimeter and sit her down by Josh. "All right, we are walking into your mom's apartment. It's so dark in here. Where is the light switch? Oh, there it is."

"Surprise!" Josh and I shout.

"Wait, don't take off your blindfold yet," I say. "Look, everyone is here. Your aunt Karen is here. Jen is here. There's Pato and Moriah. There's Ben. And, oh, your mom is bringing you a cake. Okay, here it is in front of you. Now open your eyes . . ."

She takes the blindfold off and gasps. For a few seconds, she just stares at it, as though she can't actually see it. Her face is utterly bewildered. Taped on the box sitting next to the cake are all the pictures we have of our family and friends — our "guests."

"What?!" she exclaims. "I don't — how?" She is starting to look more horrified than excited. "Did the interrogators bring this?" she asks, looking to Josh and me sharply.

"No, we asked for it from the Food Guy." I know that the interrogators approved it and that they will gloat about it later, demanding our gratitude, but I don't want to think about that now. I don't want it to be tainted, though it is.

I can see her trying to forget that the sweetness we are about to experience was given to us by our enemies. She is trying to appreciate the pictures and memories of those we love and forget about how pitiful we are, all alone here and trying desperately to be happy.

The cake is delicious. There is a layer of banana cream and walnuts. The frosting is some kind of coffee chocolate. We eat two large pieces each, smiling goofily with each bite.

59. SARAH

It's the last day of the month of Ramadan. I saw on the ticker this morning that 2,800 prisoners will be shown Islamic clemency and released. I'm happy for them. Someday that will be us.

The door opens, and Nargess and Maryam tell me to get dressed. The two of them stand there and watch as I frantically throw on my clothes. I'm thrilled to be leaving my cell for whatever reason. I already saw Josh and Shane this morning—maybe we're going to get an extra visit.

"Are we leaving the prison?" I ask, thinking perhaps Salem has come to celebrate Eid with us. Maryam arches her eyebrows and Nargess rolls her eyes. As usual, they know nothing.

"Where are the two American men?" Nargess asks the guard at the end of the hall. He picks up the receiver on his desk and talks to someone on the other end. "Only the woman," he says. "She's going alone today."

"What?" I say. "Where are Josh and Shane?"

"Only you," Nargess says angrily, grabbing my arm. "Hurry."

When we get downstairs, a group of men in suits is in the hallway. When we pass by, one of them says, "Hello, Sarah."

"Salaam," I reply. *"Eid Mubarak,"* I continue. *Happy Ramadan.*

"Eid Mubarak," he replies. "Are you happy, Sarah?"

"No," I say, accustomed to these kinds of smug, insinuating questions from faceless men in suits, "I am not happy."

"Your Farsi has become good," he replies. "You should be happy."

Nargess leads me outside. Tonight, the world's Muslim population will break its month-long fast at the end of Ramadan. It's the
second-biggest holiday of the year. The prison clinic will be closed; so
will any government office. Where can she be taking me? Whatever
this is, it's not fair that I'm out here while Shane and Josh are still in
their cells.

After walking a few minutes, we stop in front of a building I've
never seen before. I'm led up the stairs into a large conference room.
There are cameras on tripods on one side of the room. Nargess takes
me into an adjacent office, closes the door, and locks it behind me.

I tear off my blindfold and scan the room. There's no one here.
Shit, I think, will they try to coerce me into making some sort of
public confession? If they think they can single me out and bully
me, they're wrong. I'll refuse to say a word until Shane and Josh are
brought here. I walk over to the desk and rustle through a pile of paperwork, all written in Farsi. I open the desk drawer, see a pencil,
grab it, and quickly hide it, along with three paper clips, inside my
bra. I can hear the conference room filling up with more and more
voices.

About twenty minutes later, the office door opens and I'm told to
come forward. At the far end of a long table, flanked by ten to fifteen
men, is the Omani envoy we met for the second time a few weeks ago,
Dr. Salem al-Ismaily. I rush toward him.

"What's happening?" I ask. His proud face looks tired and worn,
his eyes distressed.

"Sarah," he says, "I like to think I'm a reasonable person. But I don't
understand why your Hillary can't keep quiet!" I understand implicitly that he's saying this more for the benefit of the people in the room
than for me.

"What is happening, Salem? I have no idea what's happening!"

"Sarah, you are going to deliver a speech tomorrow. You will ask
President Ahmadinejad for a pardon."

"What? A pardon? Salem, why aren't Shane and Josh here? Where
are they?"

"Sarah, I came here to bring you all home. I've been working very,
very hard on this, but many things have gone wrong. For example,

your Clinton insists on saying insulting things about the Islamic Republic at the worst possible time. You think this puts them in a giving mood?"

"Look, Salem, I can't do this. I can't leave here without Shane and Josh. You can't expect me to do this."

"Sarah." He looks at me and his eyes soften. We've been talking as if we're the only two people in the room when in fact there are dozens of eyes and three video cameras glued to us.

"Sarah," he says, lowering his voice, "you are going to help me get them out. You're going to have to trust me. I need you on the outside. We will get them out together."

I look at him strangely. A thousand thoughts race through my mind, but I drown them out. Salem's eyes are big and warm. I feel like I've known this man for a long time. It suddenly dawns on me that this feeling is the sign I've been looking for. He's part of my destiny, I think. I have to trust him.

"What do I need to do?"

"You need to write your speech, quickly. You will deliver it tomorrow, early."

"Right. Can I please have Shane and Josh help me with the speech?"

"No, Sarah. This is your speech. You will speak for them."

"Okay." I pause and close my eyes. "Give me a pen."

Salem and I work on my speech for half an hour. I know exactly what I want to say, but I'm finding it hard to battle with the voices in my head. What do Iran's leaders want me to say? How can I please my captors? What can I say to make them release us all together?

There are no magic words. For Salem, the only essential point is that I thank the Supreme Leader and President Ahmadinejad. I agree without hesitation. I will ask the government to pardon Shane, Josh, and me. I will say that we never meant to cross the unmarked border between Iran and Iraqi Kurdistan, but if we did, we're very sorry. Shane and I came to live and work in the Middle East. After we'd been there over a year—when our good friend Josh came to visit—we wanted to show him the side of the Middle East we love—far different from the violent, depressing headlines we've grown accustomed to in the United States. Ironically, Shane, Josh, and I are now the subjects of those sensational headlines, but this story can still have a happy ending. Please, I'll say, let Shane, Josh, and me go home to our families. End this misunderstanding. We've suffered enough. We meant no harm. Let us go home together.

When the speech is complete, I'm taken back to my cell.

60. SHANE

I hate it when this happens — when dinner comes and we eat and there is still no word of us going to *hava khori*. I used to go ballistic and pound on the door when they were late. I've gotten a lot more patient. They always come eventually, but it's really late now. The sun has gone down. We pressed the button a half-hour ago and no one has come. Something is up. Earlier today they brought us downstairs, stood us in front of a bunch of men in suits, and brought us back to our cell with no explanation.

"I think I want to knock," I say to Josh, whose nose is in his book. He nods.

I knock loudly, but not harshly — with just the edges of my knuckles, three raps at a time. Eventually, a guard comes.

"*Chi?*" he says.

"*Hava khori mikhaam,*" I say.

"*Neestish,*" he says. No *hava khori.* Josh jumps up and we crowd the door. The guard walks away as we start to argue with him.

This time I use the butt of my fist and pound the door hard. He comes back and tells us that *hava khori* is out of service. The door is broken and they are fixing it. I feel that thing happening — where they tell a lazy lie and I start to believe it, because if I believe it, I don't have to be worried. But Josh is better at not playing that game. He insists that we must go.

The guard leaves. We wait five minutes. We pound on the door again. He returns. "Sarah is sick," he says this time. "She can't go outside."

"I want to see her anyway," I say.

"You can't. She says she doesn't want to see you." He says it as if he were saying, "Fuck you."

We don't give up and eventually the warden comes. It's not going to happen, he says. We are not going to see Sarah. If we want to go outside, the two of us can go to one of the small open-air rooms. But no Sarah. Inside, I'm in tumult and my body wants to keep going, to make a scene. But some part of me somehow knows that something good is happening. It's *Eid*.

61. SARAH

I spend the next fifteen hours pacing my cell and performing my speech again and again. By 3 a.m., I can go through it ten times without a single mistake. The dots of sky I can see through the perforated metal over the windows are turning a dark gray. As the sun comes up on September 11, 2010, it hits me that the Iranian government chose this day, of all days, to make a statement. We're more benevolent than America? We held this woman as revenge? Even I don't understand what the message is supposed to be.

When the breakfast tray comes, I jump up and prepare to go. I have no idea when they're coming. All I can eat is a few bites of bread before I hear Shane and Josh whistling from *hava khori*. They must have seen the ticker, "American Woman to Be Pardoned by President." What do they think? I hear them whistle the melody of Bob Dylan's "Lonesome Death of Hattie Carroll." We agreed a long time ago that that song would be our code in case we were separated. It means Shane and Josh are hunger-striking.

At some point during the night I've completely accepted my fate. I've never felt so confident, ready, and clear. Shane and Josh have kept me alive the last fourteen months; without their love and selfless support I would have lost my mind and probably hurt myself many times by now. Now it's my turn to return the favor by getting them out of this prison.

I turn on the television, crouch down, and begin scanning the ticker at the bottom of the screen, as has been my habit every morning for nearly a year. A headline, seven words in length, crosses the screen, then disappears. I freeze. I don't move an inch for the next five minutes, waiting for the headline I think I saw to come back after the others cycle through. Did I imagine it? No, there it is again. It's real.

"Pardon of American Spy Canceled by Judiciary," I read again.

The pain lasts only a few seconds. I've gotten very good at dealing with disappointment. Actually, a part of me feels relief. At least I'm going to see Shane and Josh soon, I think. We'll get back to the old routine. I ball up my speech and throw it in the trash. Then, the door opens.

"Sarah, get dressed."

I'm led outside our building, into sunlight, and along paved roads. We pass the building with the conference room where I wrote my speech yesterday. We walk by the locked gates leading out of Evin Prison, but instead of leaving the compound, we turn right and enter a small building.

Inside, I'm taken into an office where there are four or five men seated around the room. I hear the voice of Dumb Guy.

"Sarah, we have good news. Very good news."

"What are you talking about? Didn't you see the news? My pardon has been canceled."

"No, Sarah, you are going home. You just have to do it the way the law says. You have to sign this paper."

"I'm not signing anything. Where is Salem? Where's my lawyer?"

"Sarah, don't you want to go home? You have to trust us," Dumb Guy whines.

"I will never trust any of you." I lean back in my chair. "If you want me to sign something, you'll have to bring my lawyer here first."

"Sarah, this man is the judge — you don't want to offend him." It's Father Guy's voice — I didn't even know he was here. "Sarah," he says, "you must admit that I have never lied to you. If you have any trust left for me, please believe me now. If you sign this paper, you will be released. You will go home."

I cross my arms and say nothing. Is it true that he's never lied? I decide it is, to the best of my knowledge. What does it matter? I don't know if it makes sense anymore to leave here without Shane and Josh — I don't know what signing this paper might mean. I resign myself to do nothing.

"Okay, Sarah," the "judge" says, "we will call your lawyer, but we don't know if he will be able to come on such short notice."

In only fifteen minutes a man walks in the door. He greets each man in the room one by one, shaking their hands and kissing them on the cheeks. Then, he turns to me.

"Hello, Sarah. I am very happy to meet you. I am Masoud Shafii, your lawyer."

"Can I take my blindfold off?" I ask. There's no immediate answer, which I take as a yes and tear it off. I scrutinize the clean-shaven face, receding hairline, kind eyes, and nervous smile of the man standing in front of me.

"How do I know you're really my lawyer?" I ask. Indignant murmuring fills the room.

"No, please," the man says, "Sarah is right to ask." He shows me his ID and business card that reads "Masoud Shafii, Attorney." Below his name is the phone number I memorized nine months ago when we signed a paper accepting this unknown man as our lawyer. He sets his briefcase on an empty desk next to me, opens it, and removes a document. I can see our signatures are at the bottom next to those of our mothers, Nora Shourd, Laura Fattal, and Cindy Hickey.

"Okay," I say, "so you're Masoud Shafii. What am I supposed to do?"

"I advise you to sign the paper," he says.

"What does it say?"

He sits next to me and does his best, in halting English, to translate the document line by line. The paper is an indictment for two charges — illegal entry and espionage.

"So, after almost fourteen months in prison, you're charging me with espionage and then releasing me?" I ask.

"Yes, Sarah, exactly," Dumb Guy says.

"What does this mean for Shane and Josh?" I say, ignoring him and looking at my lawyer.

"It means, well, hopefully they will meet you soon," Mr. Shafii answers.

"What do our parents want me to do?"

"They want you to sign the paper."

"What does Salem want? The Omani envoy?"

Mr. Shafii looks confused, talks to the others in Farsi for a few minutes, and says, "If you sign, he will be here to take you home soon."

I realize it's time to end this scene. If I've surrendered a thousand times before, I can do it now. I pick up the pen and steer it toward the page.

"Wait," I say. I turn and look Father Guy straight in the eyes for the first time. All those hours I had sat in the interrogation room with him; I had never really seen his face.

"Promise me I won't leave here without saying goodbye to Shane and Josh," I say.

"I promise, Sarah."

"Okay." I sign the paper.

62. SARAH

In Greek mythology, the underworld is not an easy place to leave. Hercules had to cross the five rivers, pass through the adamantine gate guarded by Cerberus, the three-headed dog, and cross on the single ferry steered by Charon, who demands a single coin.

It soon becomes clear what my payment is. They take me to a garden inside the prison complex where a man and video camera are standing next to a fountain, waiting for me. He tells me he's from Press TV, an Iranian state-run English-language satellite network.

"How do you feel to be free?" he asks.

"I'm still in prison, aren't I?"

"Can you show us your ring?"

"Why am I being released without my friend and fiancé?"

"What was it like for you in prison?"

"It was terrible."

Then, they lead me up some steps to a room with two couches facing each other. I sit on the couch in silence and try to avoid looking at the cameras.

"What am I doing here?" I ask the cameraman.

"We want you to meet someone," Dumb Guy answers.

Two teenage girls — they look like twins — and an old woman and an old man walk into the room and awkwardly sit on the couch across from me.

"We are happy for you to be free, Sarah," the old woman begins, "but your government should free my innocent daughter."

"Who is your daughter?" I ask.

"Shahrazad Mir Gholikhan," she answers.

"Yes, of course, I've heard about her on TV."

"You have a TV?" one of the girls asks. She speaks perfect English and her voice has a twang, almost a Valley Girl–type affectation. "Our mother's food is dirty — it has bugs in it. How is your food?"

"It's okay," I reply. "Where is your mother being held? Do you talk to her?"

"Yes, sometimes, and we get letters."

"I've never been allowed to write a letter. I've only been allowed two five-minute phone calls in thirteen and a half months."

"Really?" They look surprised. "Ah, well, we want you to ask President Obama to free our mother. We miss her. We are very sad without her. We wrote this letter for you to give him." She hands me a piece of stationery covered with large, voluptuous handwriting.

"Okay, I will do what I can."

The conversation seems stilted, scripted. I feel sorry for these girls, but I know nothing about their mother or why she's in prison. A part of me wonders if the story is true. It might be true, I think, but do they really think I can do anything about it?

63. SHANE

A guard takes us out of our cells. We've been on hunger strike the last twenty-four hours, threatening that we won't stop until we see Sarah. On the way to *hava khori*, two guards stop us. One tells Josh to stand against the wall and the other brings me into an interrogation room. He is wearing rubber gloves. Without preamble, he pulls down my pants and tells me to pull down my underwear. He pats my legs from my ankles to my thighs, as one would do if the person they were searching were wearing pants. From between my legs, he looks up under my penis without touching it.

Josh goes next.

When we step outside, Sarah is sitting down. She looks calm and ready. Josh and I sit down and each take one of her hands. "You guys, as soon as I get out, I'm going to fight as hard as I can until you are free. I will not stop. There will be no celebrating until you are out — not even one drink." She tells us everything that has happened in the last couple days. It is clear that she believes in Salem. She is sure that he will get us out.

I have no idea when I will see her again, but I feel a deep relief starting to settle in. She won't be alone anymore. She'll be out and she'll have a purpose. She'll stop withering away and start regenerating. What will it mean to miss her? I can never want her to be with me here. Missing her will just become part of my yearning for freedom.

I can see that she feels guilty. Part of her feels like she is turning against us by leaving us behind. She isn't sure if she is doing the right thing. Josh sees it too and tells her how happy he is that she is getting free. We tell her over and over, and each time, her face looks a little less burdened.

"How long do you think I should wait in Oman for you?" This question echoes like my mom's hope that we would be coming home with them. It hurts a little, because I know that she can't accept that we will probably be here for a while. I remember my mom's face as that elevator door closed. "Sarah," I say, "I don't think they are going to let us out soon. But that's okay. We are going to be okay." Maybe I'm trying to reassure myself more than her.

"You guys have to make sure you take care of each other. I worry about you. Be open with each other. Don't just read books all the time. You have to talk about your feelings."

"You need to take care of yourself too, baby," I say.

She scoffs. "Yeah, okay. I'm going to take care of myself, but I'm going to get you free!"

Josh and I start to tell her things that we've never told her about our lives in our cell. Until now, it has mostly remained a secret world that we've kept private to keep her from feeling jealous. We tell her that we give each other massages and that when we wake, we discuss our dreams.

Sarah says she is going to study Arabic every day, and math for the GRE. I tell her to make music, to use her music to help get us free.

The door opens. Unlike almost every other time, we don't protest today. We don't ask for more time. The guard doesn't push us either.

"I feel like I'm one-third free," Josh says.

I don't know what the appropriate thing to do is in this moment. I kiss her and hug her. I tell her I can't wait to marry her. I smile. I don't know when I will see her again, but I know it will happen someday.

"Goodbye," I say. "You are the best person to fight for us. I love you, baby." Letting go of something I love has never felt so right.

The guards walk us down the hall, she turns off to her cell, and Josh and I keep walking. Unlike almost every other time I've left her at that junction where we split from each other every day, this time I don't look back.

64. SARAH

"Don't forget me in America," Maryam says, helping me get dressed. I'm back in my cell after a brief goodbye with Shane and Josh in *hava khori*. A plastic bag has been placed on my bed. In it are a hairbrush, a new hijab, bright white tennis shoes, jeans, and a fancy-looking *manto*.

I feel even more confident after seeing them. Now that I have a mission, nothing else matters. Leila shows up. I take her hands and beg her to look out for Josh and Shane from now on, not to let the male guards be violent with them. She agrees, telling me how pretty I look, how happy she is for me. "Don't worry," she says, "this is going to be okay. You will be married soon!"

"*Inshallah,*" I say to Leila.

"*Inshallah,*" she repeats.

I look back at the long, white hallway in which I've spent more than a year of my life. No more blindfolds, no more *hava khori*. These women are my friends, I realize, watching Maryam fuss over the pile of things I've left for Josh and Shane while Leila stands with her hands on her hips, giving me motherly advice. We've been through so much together; I can't help but care about them. They lead me down the hallway with my backpack swinging over my shoulder. At the end of the hall I pause and turn back, knowing that behind every door there is a prisoner crouched by the slot listening to us — just like I have done so many times.

"*Inshallah azadi!*" I shout down the hall, feeling the first jolt of excitement shoot through my veins. *God willing, freedom!*

"*Inshallah shoma azadi hameeshe!*" I shout louder. *God willing you freedom forever!*

Leila and Maryam usher me away, but I turn back, peering under my blindfold to look down the hall one last time. I hear a clap, quiet and timid, coming from inside one of the cells.

"I love you!" I shout — my last words in Section 209.

Like a zombie I follow Nargess into a small room where she makes me strip, one last time. I don't look at her; I just wait patiently and stare at the ceiling as she looks between my legs, swipes her hands under my breasts, and peers in my ears.

I'm then taken to another building, where male prisoners are lined up without blindfolds. A few of them recognize me, point; then one of them waves. Are they getting free too? I'm led past them to a booth where a woman takes my fingers one by one, coats them in ink, and presses them to a sheet of paper. A car with tinted windows is parked outside. An older, conservatively dressed woman waits in the car. As we leave the gates of Evin Prison, she pulls my head down on her lap and smothers me in the folds of her black, flowing garments. It's not an act of affection — she's there to make sure that no one will see me as we speed through the streets of Tehran.

I leave Evin Prison in darkness. I hear the huge gates shut behind us and feel the car pick up speed. I close my eyes and relax into the lap of this unknown woman, not resisting, not even stressed. If Shane and Josh were here, we'd be laughing and crying for joy right now. Instead, I feel my body preparing for the battle ahead, using these few moments to preserve my energy and gather my thoughts.

I don't hear any other cars on the freeway — is it blocked off? The car slows and the woman uncovers my face. She smiles down at me and tells me to wait. I sit up and I'm led inside a small building. Dumb Guy tells me we're at the airport. We enter another formal, furnished room, with glass and mahogany tables and plush satin couches. I scan the room for Salem, perhaps Ambassador Leu Agosti, but they aren't here.

Cameras arrive. Iranian reporters hook small microphones to my clothes and ask me how I feel. I feel the way a cat feels, I think, when you pick it up by the loose skin on the back of its neck. I feel limp, out of my body, unable to resist.

"I feel grateful and humbled by this moment," I say. "I want to offer my thanks to everyone in the world, all of the governments, all of the people that have been involved, and I particularly want to address

President Ahmadinejad and all of the Iranian officials, the religious leaders, and thank them for this humanitarian gesture." I pause, trying to remember my well-rehearsed speech. Up until this point, the words have been rote — it's what I've been told to say. I let myself speak from my heart.

"I have a huge debt to repay the world for what it's done for me. My first priority is to help my fiancé, Shane, and my friend, Josh, to regain their freedom, because they don't deserve to be here. Even when that's finished, I feel like my work has just begun repaying the world for what it's done for me. I realize that there are many innocent people in prison that don't have the kind of support that I've had. I feel humbled and grateful and ready to be free in the world again and to give back what's been given to me."

The questioning and photographing go on for two and a half hours until I can't take it anymore and beg them to leave off. I'm used to this game by now, used to being manipulated and controlled. But I'm already beginning to sense that I have more power than I did a few hours ago. I walk over to a couch in the corner of the room and lie down. I hug a pillow to my chest, close my eyes, and pretend to sleep. Moments later, I hear soft footsteps approach. I open my eyes and find a camera poised a few feet from my face.

An hour later, Salem is sitting across from me in his small private plane, looking harried and impatient. I ask him if everything is okay.

"Not until we're in the air, my dear. Things have been known to go wrong, even at the very last minute." Ten minutes later we're in the air. I'm trying to make out Evin Prison as we fly over Tehran. I imagine Shane and Josh in their cell, talking and planning intensely together. For what feels like the thousandth time, I thank God they have each other. I turn to Salem.

It's the first time I've sat across a table with anyone for over a year. He tells me that my mother and uncle will be waiting for me at the airport in Muscat, where they've been for several days. We're served steak and potatoes by waiters dressed in Omani garb. I eat mechanically, reminding myself that I'll need my strength. Then I excuse myself and walk down the aisle to the lavatory.

In a mirror flanked by bottles of perfume, I study my pale, angular face. I have the feeling that Salem, a man I barely know, will now be my closest ally and friend. I take off my hijab and cringe at the sight of my thin, dry hair. I put it back on, pinching my cheeks to add some color and spraying myself with perfume.

I walk back to the table and sit across from Salem. "How do you feel, Sarah?" I look out the window at the city of Muscat, thousands of feet below us. Everyone's been asking me this, but for the first time I have enough respect for this man to reply honestly.

"Nothing. I feel nothing, Salem."

"It's okay. That will change," he says.

The feeling doesn't change as we descend, nor when we bump to a gentle landing. It doesn't even change as I walk down the stairs and touch free ground. At first, I can't see anything but cameras flashing. As my eyes adjust to the darkness, a crowd of people materializes.

I scan the crowd anxiously for my mother. And there she is. I walk toward her and I see her face, so strong and vulnerable at the same time. I see her tears. I wrap my arms around her and everything else drops away. My body belongs to me again. Though we're surrounded by reporters, in that moment it's just she and I. We're finally alone.

I feel the cold air on my bare head and reflexively reach up to adjust my hijab, wanting to cover my thin, brittle hair from the cameras. *No shame*, I tell myself, letting it drop. I look into my mother's eyes and smile.

65. SARAH

My eyes spring open. The dark sky outside the tall windows is already showing hints of blue. After talking through most of the night, I fell asleep with my body half draped over my mom's torso, my head resting on her outstretched arm. I quietly get up, pause to marvel at my mom's sweet, tired face, grab her cell phone, and tiptoe downstairs.

The embassy maids are already bustling in the kitchen. "Sarah," a Filipino woman in her forties greets me with her arms stretched open, "we were so worried about you." She hands me a mug of steaming coffee and shows me a close-up picture of my face on the cover of the *Muscat Daily News*. I pause to inhale the coffee's intoxicating smell — one of the countless things that's been cut out of my life. I open my mouth as if to ask permission, then think better of it — and simply walk outside.

It's a little after 5 a.m. and I'm standing on the edge of a white beach with warm turquoise waves lapping at my bare feet. The air smells like sun and salt. There are men and women, the former in long, white robes and the latter in black, already out for a morning walk along the shore. Behind me is Embassy Row, dozens of three- and four-story pink and beige mansions ringed with opulent flowers and flanked by tropical trees. The calm, placid ocean stretches out for miles and miles before me, as if there were no end to the earth at all.

Talking to my mom last night, I felt alive and exhilarated. Now, looking at the sun reflecting like daggers off the delicate waves, it hits me just how heavy a toll prison has taken on my imagination. No matter how hard I tried, I could never evoke this kind of beauty in my mind.

I place my empty mug by a rock at the foot of a palm tree and begin to scroll through the contacts on my mom's phone. I've known for a long time who the first person I called would be. I find the number, stand up, and turn to nod at a tall, beefy former marine sitting on a bench twenty feet behind me. He gives me a wave in return. In the last twenty-four hours I've traded prison guards for bodyguards. Richard Schmierer, the American ambassador to Oman, whose residence we're staying in, warned me that international journalists are stalking me outside the embassy. He says I need protection. I start to walk down the beach and the former marine follows behind at a respectful distance.

I dial the number.

"Shon, it's me," I say.

"Sarah? Oh, I'm so relieved! You have no idea how worried I've been about you guys. I'm so glad you're out, so glad you called. How are you?"

"I'm okay. I never thought it would happen like this, Shon."

"I know. It's so awful, Sarah, but the campaign is going to be much, much stronger with you in it. This is crazy. I can't wait to see you. Is there anything you need?"

"Yes, I need to tell you something. Shon, we were so glad you weren't with us that morning. You saved our lives, calling the American embassy in Baghdad and our families. If you hadn't stayed in the hotel when we left for Ahmed Awa that night, no one would have even known we were captured. We might never have been heard from again."

Shon tells me how terrible it was not knowing what the Iranian government was going to do with us. He tells me about an American ex–FBI agent, Robert Levinson (later revealed to have been contracted by the CIA), who disappeared on Iran's Kish Island almost five years ago without any word from him since. "Iran doesn't admit to holding him," Shon says. "No one knows where he is. He might even be dead."

"It's so strange to feel so lucky but so unlucky at the same time," I tell him.

"Are you really okay? How's your health?"

A few weeks ago, the prison doctor called me into his office to show me the results of my mammogram, proof that the lump in my breast was benign. Even though they appeared confident I didn't have cancer, some Iranian officials still pointed to my "medical condition" as the reason for my release — and the international media ran with it. On the day of my release, when I asked the judge why I was being allowed to go before Josh and Shane, he told me the decision was based on my gender and solitary confinement. The timing, however, was clearly calculated to ease international pressure leading up to the UN General Assembly meeting in New York so they would not appear to be giving in to the United States. I credit this strategy largely to our campaign. By allowing my mom to push my "health problems" and isolation in the media, they gave the Iranian government a way out.

"I'm fine, Shon. You and our other good friends have been working closely with the families, right?"

"Yeah, it's been amazing. Your mom, Cindy, and I are so close. But there's division around strategy, and sometimes it gets personal. People don't agree on what should be done."

"Well, it should be a lot more clear now that I'm out. We can find out what worked in my case. Or didn't work," I say.

I hang up with Shon and start to jog down the beach. Before I can do anything or decide anything, I need to feel like myself again.

Shon's words have shaken me up a bit. I'm worried about the conflict he referred to around the campaign's strategy, but I know some exercise and a good breakfast will help me stay grounded. I'll eat eggs. Eggs! Then, I'll find someone to cut my hair short, something I was never allowed to do in prison. I'll talk to Shane's and Josh's family members, one by one, and then I'll talk to my own. Later, I'll meet with Salem, and we'll begin to devise a plan.

A few hours later I'm back on the beach. I'm spending as much time as possible outside, since feeling the fresh breeze and sun on my skin makes me feel one thousand times more alive. Salem is walking beside me, wearing a long white robe and carrying a thin, ornate cane. I'm wearing new clothes that I bought today at a local mall, black cotton slacks and a long, white V-neck.

"Your haircut looks lovely, Sarah, but where are your shoes?" Salem asks.

"I'm sorry," I say, suddenly embarrassed. "I like to walk barefoot on the beach."

"Sarah, are you a hippie?" Salem asks with a sparkle in his eyes.

I laugh. "If walking barefoot on the beach makes me a hippie, then I guess I am." I turn serious. "Salem, I have so many questions to ask you."

"Ask me anything you like, my dear."

"Who paid my bail? Was it you?"

"Sarah, Omanis are not in the habit of asking for credit for what we do — it's not our culture."

"That's incredibly generous of you, Salem, but the media is going to ask me who provided half a million dollars in cash — what should I tell them?"

"Tell them the truth, that all you know is that you were brought out of Iran in an Omani plane, that we led the negotiations and will continue to do so for Shane and Josh until they are freed."

"*You* led them?"

"Yes, of course. Do you think I could leave Iran without you? But you must remember that everything I do is for His Majesty Sultan Qaboos bin Said."

"How are we going to get them out, Salem?"

"You and I are going to work side by side. You must show Iran's

president and Supreme Leader that they didn't make a mistake releasing you. You must show them that you are fair, that the boys will
also be fair."

"Okay, I can do that. We're going to have a press conference in
New York in three days. Then next week I go on *Oprah*."

"Perfect, Sarah, everyone loves *Oprah*. My wife and daughter will
be very happy."

I explain that we chose *Oprah* because it's the most watched
American show in the Middle East. I ask Salem how he got involved
in our case and he tells me he was first approached by an Iranian family residing in the United States. From the start, this family knew we
were innocent and that our detention would only serve to worsen the
animosity between the countries. Three months into our captivity,
they decided to approach His Majesty Sultan Qaboos, with whom
this family has old ties, to serve as a behind-the-scenes intermedi-
ary with the goal of quickly resolving our case. The sultan agreed, ap-
pointing Salem al-Ismaily as his envoy.

"Do the families know about this Iranian family?" I ask.

"No, no one knows about this."

"I heard that several members of our families approached the
Omani embassy early on, asking for your government to mediate."

"Yes, that's true," Salem answers. "That was very helpful. It encouraged your government to officially ask His Majesty to get involved — from that point on we made your case a priority."

"So, what happened next?"

"The tricky part was to get to the Supreme Leader," Salem goes
on to tell me. He says that with the help of friends, he was granted a
meeting. He brought something special to this meeting, a Quran that
he made for his daughter after she lost her sight. On every page there
is a button to press, and a recorded voice reads the Surahs aloud. Salem showed this to the Supreme Leader and apparently he was very
impressed. "Then," Salem continues, "I mentioned your case was a
special consideration for the sultan. It made an impression."

"But why did the Supreme Leader agree to my release but not
Josh's and Shane's?"

"It wasn't the Supreme Leader — it was the judiciary. There are
disputes between the judiciary and the president's office. The Su-

preme Leader has already approved their release, but he says the president and the judiciary must sort this out among themselves for the good of Iran. Ahmadinejad was going to pardon you all, perhaps even bring you to New York on his own private plane, but the judiciary managed to stop him."

"So, the Supreme Leader has agreed to release Shane and Josh?"

"In principle, yes, but all sides must agree. The judiciary, the parliament . . . We cannot risk another mess like what just happened."

"How are we going to do that?" I ask.

"I will meet you in DC and tell you more soon. For now, focus on your speech, Sarah. Focus on *Oprah* and, please, buy yourself some shoes."

"Okay, I will, Salem." I smile and reach out to shake his hand. I'm reassured by Salem's brilliant diplomacy—his capacity for truly understanding the frustration on both sides. After walking along the beach for hours, we've made our way back to the embassy. "Salem, I am really going to enjoy working with you. I can never repay you for what you've done for me and my family."

"Sarah, you can repay me by helping me finish this job. Then, we can plan your wedding."

Back upstairs, I climb a small ladder onto the roof of this mansion, needing to see the ocean one more time before I go downstairs for dinner. I've spent the whole day talking, but I feel like I've only scratched the surface of what I need to know. I take a deep breath, close my eyes, and open them again. The palm trees are swaying with the breeze and the setting sun is reflecting off the waves like a million bright sparks, as if a giant blaze were hidden right below the horizon. The sight brings the familiar sting of tears behind my eyes. I truly forgot the world was this beautiful.

66. SARAH

When I step off the plane in Dubai to catch my connecting flight from Oman, there are FBI agents waiting for me. They ask me to follow them into a backroom to answer some questions, and then start showing me pictures of Iranian men, asking if any of them looks familiar. They ask me if I know anything about Robert Levinson, the missing ex–FBI agent. I tell them, politely, that I'm not ready to talk to them, but that they should contact me in a few weeks when I've had time to gather my thoughts.

As soon as my mom and I step on the plane to the United States, a flight attendant informs us that we've been bumped up from first class to luxury class. They bring me a vegetarian meal and warn me that there are people on the plane, most likely journalists, who know who I am and are trying to talk to me. A few minutes later a man I don't know walks up to me, smiles, and hands me a small box. There's a note on the top: "From a fellow American citizen with love."

Inside the box is a duty-free watch. I try to imagine what I symbolize for this man. American resilience? National pride? I'm not sure that I've ever felt anything akin to nationalism in my life — my core identity has largely been shaped in response to (and rebellion against) what I dislike about my culture . . . greed, selfish individualism, a sense of superiority and entitlement over others — but the connection I feel to this man is undeniable. Thanks to the Iranian government, I've never felt so American in my life.

The flight attendant apologizes — she tells me they're trying to keep passengers away, but it's difficult. I turn on my personal TV and flip through the channels until I find the news. After a few minutes, an interview with ABC's Christiane Amanpour and Ahmadinejad comes on.

"We let Sarah go," he tells her, referring to me by my first name, as if we're old friends. "You may be aware that eight Iranians are being illegally detained in the United States, so I believe it would not be misplaced to ask that the U.S. government should make a humanitarian gesture to release Iranians who were illegally arrested and detained here."

"Are you saying you're holding the two Americans as hostages for the release of Iranians here?" Amanpour asks. Asking for a prisoner swap when they know we're innocent, before Shane and Josh have been tried or convicted, is hostage taking in no uncertain terms. Still, the president demurs.

"No, but how would you know that those Iranians are criminals?" the president retorts, raising his eyebrows and tilting his head. "Are you a judge?"

Disgusted, I turn the TV off. At least they are calling us hostages on mainstream TV, I think. It's much more accurate than "hikers." Being called a hiker is kind of like being called an omelet maker — it's something I enjoy doing, but by no means is it a core part of my identity. It also doesn't offer any explanation as to what brought us to the part of the world where we were captured. I press a button and my seat transforms into a small private pod. I lie down, drape a blanket over its sides, and take out the speech I've written for the press conference I'll give in New York a few hours after we land.

As I fall asleep, I think about talking to my sister, Martha, on the phone that morning. "Sar," she said, "I've been worrying about you every second of every day." The interrogators never allowed me one letter from my sister, telling me that because we don't share the same father, she is not considered to be immediate family. As much as I tried to resist it, not hearing from her for over a year had forced me to stop thinking about her, and a part of me had even begun to believe that she didn't care. "You are so important to me, Sar. You're my only sister and I need you." Her words were like balm; they found a way through the hard shell I had around me, and my first free tears began to flow.

I hear the flight attendant calling my name. I open my eyes and she tells me that the pilot has announced our descent. I'm finally back in the United States, but I feel like I'm leaving a big part of who I am behind me — and right now Shane and Josh are the only people in the world who understand me; even my own country feels foreign.

When we get off the plane, Josh's brother, Alex, is standing in the lobby along with a few people from the State Department. I'd seen pictures of him in prison, but it's not until this moment that I realize just how much he looks like Josh. I can't look away.

"The room will be full of cameras," Alex says as he, my mom, and I get into a private car bound for New York. "The best advice I've heard is to focus on a point on the wall above and behind them."

I interrupt him. "Alex, I'm really sorry that Josh is still there. We had no idea this would happen. You can ask me anything, anything at all."

"Have they hurt him" — his upper lip twitches — "physically?"

"No," I say, "it's all psychological, but we both know how resilient Josh is. He's going to come out okay."

The only time I saw Josh cry in prison, his tears were in sympathy with my tears. I never saw him cry for himself. Now, with my arms wrapped around Alex's shoulders, it's as though I'm with Josh again, finally seeing his tears, as if I'd never left him at all.

67. JOSH

Dumb Guy shows up at *hava khori*. He makes sure we've received all the books from Sarah's cell. Then he makes it clear he won't come back for at least two months, and I get peeved. That means no letters, no phone calls, no nothing for two months.

"We can't go that long without contact," I tell Dumb Guy. "Is this how it's going to change without Sarah? Why won't you come sooner? What's wrong? Do you miss Sarah already?"

Dumb Guy stares at me blankly, then makes his way to the door. Before leaving, he says dryly, "I think *you* miss Sarah." I don't argue.

Back in the cell, I stare up at the window and think about her. I think about all the freedom dreams we hatched together. The three of us will hike in the woods, we'll go whale watching. We'll gather at a rustic cabin and get drunk. And we'll listen to a full album of music with our eyes closed.

When I'm eighty years old, what difference will it make whether this ordeal takes one, two, or even three years? For me, the most difficult part about imagining a longer detention was the prospect of Sarah struggling through it in solitary confinement. With her free, I feel more relaxed.

I notice the moon at night signaling the approaching Jewish New Year. But I'm not motivated to celebrate Rosh Hashanah without Sarah. I won't celebrate the holidays anymore. I realize how much she encouraged me. For her, I strived to honestly say, "I'm doing fine," and to be able to be there for her. I didn't realize that her simple "How are you?"— so full of empathy — meant so much to me. I didn't realize that I gained a sense of purpose from caring for Sarah.

It's 11 p.m., and Shane and I are watching the nightly English news, waiting for a story about Sarah. For the past few nights, we've watched attentively and haven't seen anything. There *has* to be a story about her.

All I see is the back of her head, but I recognize her purple hijab immediately. She walks slowly toward an airplane. Then the screen cuts to her at a press conference speaking to the cameras.

"I want to thank Ayatollah Sayyid Ali Khamenei for his compassion." Her voice goes silent.

"Bullshit!" Shane blurts out. "Of course, they'd only let us hear that one line."

The canned audio cuts to the commentators who laud the Iranian government's mercy. They cite "medical considerations" for her release.

"Salem's making her play that diplomatic game," I point out to calm him. Then I interject, "She looks great!"

The coverage is agonizingly short. The news turns to Ahmadinejad at the United Nations, and my heart sinks.

I worry about her. I know that leaving solitary is a huge adjustment — even within prison. I worry she'll have a hard time shedding the anxiety and anger that prisons breed. But hearing her talk at *hava khori* about Salem and the speech she memorized last week made me glad that she's the one they released. If I had to choose one advocate out of the three of us, it'd be her.

In that short glance at her face on TV, like in those first glances when she used to enter *hava khori,* I can read her emotional state. I can see her poise in public and her clarity of purpose. For the Iranian government, Sarah's freedom perhaps lessens the urgency to release Shane and me. But don't they realize how effectively she'll advocate for our release?

As long as I think about Sarah uncaged, I feel more free myself.

68. SARAH

"I feel bruised but unbroken," I say, standing behind a podium in a room filled with cameras at my press conference in New York City. Immediately after, I'm ushered out of the building, past flashing cameras to a cab waiting outside.

"Sarah," Paul Holmes says as I'm about to step into the cab, "you know, there's no reason to go straight back to the hotel . . ." He adds with a lilting British accent, "Perhaps you'd like to *walk* back?"

Paul was referred to our families by a friend. He helped them write their first press release less than a week after we were captured — weaving together three entirely different texts into a version everyone was comfortable with. Paul currently works as a communications professional — before which he was an international journalist who worked on and off in the Middle East for almost two decades. He told me he thought that editing that initial press release would be the end of his involvement. I'm glad it wasn't. Since then, there have been dozens more press releases, letters, and carefully written media talking points. Paul's become far more than the campaign's pro bono communications professional; he's the glue that holds our campaign together — actually, the *families* together. From what I gather, without Paul's calm demeanor and reasonable facilitation, internal disputes might have torn our families apart months ago.

"So, how do you feel, Sarah?" Paul asks. It's the day after the press conference. Paul and I are sitting in his cozy apartment on the Upper West Side of Manhattan. He's been gently quizzing me for the last fifteen minutes in preparation for *Oprah*.

"I feel determined," I say with hesitation. For some reason, this is the hardest question I've been asked. "Yes, determined."

"Okay, I don't doubt that. But 'determined' is one step removed. What do you feel under that?"

"Um, devastated?" I offer. "No, that's not quite it. Sad?"

"Something's missing, Sarah. If I don't believe you, the audience won't either."

"The truth is, I feel numb, Paul. I'm trying to find the rawest emotion, but I just don't feel anything."

"That's okay, Sarah. That's exactly what you should say. If she asks you how you feel, tell her you feel numb."

"Really, I can say that?"

The next day, I'm walking across an ocean of beige and pink on Oprah's stage. Trying to ignore the bright lights, applause, and rows of smiling faces, I find Oprah's eyes, anchor myself to them, and sit down. Oprah's direct, confident gaze puts me at ease. She feels like a pillar of strength that I can draw from.

"Sarah, how do you feel?" she asks.

"Honestly, Oprah, I feel numb. This is all so shocking, I can't really feel anything."

I'm not used to the heavy makeup on my face. I'm not used to the hundreds of eyes fixed on me. Yet, as I begin to speak, something about this situation feels familiar and almost second nature. I know the millions of people listening have no idea who I am and no real conception of what I've been through, but I'm used to that. I'm used to trying to translate my experience to guards and interrogators with only modest success. I'm used to being misunderstood.

I tell her that Iraqi Kurdistan is an autonomous zone, practically its own country. Unlike most of Iraq, Kurdistan is not a war zone, and no Americans have been killed or kidnapped there in recent decades.

"Tell us about Shane and Josh," she says. "Are they okay?"

"They're very isolated," I say. "This has taken a heavy toll on them. I don't even know when I'll be allowed to talk to them again . . . This is the worst separation yet." I pause and weigh my words, thinking that Salem, even President Ahmadinejad, must be listening. I decide on a lighter version of what I want to say: "We're being used."

During the break, Oprah takes off her colossal heels so we can pose for a photo at the same height. She jokes to the audience about the importance of lip gloss, referring to the story I divulged during the show about putting strawberry jelly on my lips before my "date nights" with Shane. I tell her I need to talk with her privately after the show and she agrees. An hour later, in the Green Room, Oprah walks in with her entourage. Framed by the doorway, she towers over me. I look up and tell her that I've always admired her, that we even read *The Color Purple* in prison (she was in the film version). Then I get down to business.

"I need you to help me get a meeting with President Obama," I say.

"I don't do that," she replies with iron certainty.

"I need you to make an exception," I say, standing up straight and locking my eyes on hers.

"What do you need him to know? Perhaps I can pass on a message."

"No," I say, "what I have to say is confidential . . . and important."

"Okay," she replies, weighing her response, "I can appreciate that. I'll make an exception this time. I can't promise you anything, but

I'll see what I can do." I realize that holding my ground, something I had to master in prison, is also going to prove important on the outside.

The next morning, I'm back in New York. It's 5:30 a.m. and I'm being ushered into a black car for the first of a dozen interviews we have lined up all day. Shane's mom, Cindy, is sitting next to me in the back seat. My hand is clasped in hers like an iron bolt.

I walk through the day in a haze. Between interviews, strangers come up to me on the street with tears in their eyes, asking me to pose with them for pictures. When I walk into BBC's studio, the entire room stands up behind their desks to applaud me. Having my pain, which has for so long been so private, publicly recognized feels good — but the attention is hard to process. Sustaining eye contact is difficult, almost painful, and having all these people I don't know constantly touching and hugging me makes me want to retreat into a hole. I'm finally back in the world of people — but after 410 days alone in a cell, what I crave the most is solitude.

It's also confusing when I find out how little people seem to know about us, how we are often portrayed as "young hikers,'" with pictures of Josh and Shane as children flashing on the screens behind me. One media outlet after another asks me about my romance with Shane in prison. How did he propose to you? Can we see the ring he wove out of string? Talking about Shane hurts — it doesn't feel good to have our relationship condensed to a sound bite and, to be honest, it feels cheap and almost disrespectful for so much attention to be focused on our engagement when Shane and Josh are still in prison, with no guarantee of an end in sight.

Again and again, journalists ask me if I'll go back to Iran for the trial. "I'm not ruling anything out," I say. "I'd go back to prove that none of us have done anything wrong." We came up with those lines in strategy meetings, keeping in mind that the last thing we want is for the Iranian government to get into this even deeper by having a trial, which would inevitably lead to a conviction. The answer I give again and again is strategic, circumspect. If I'm actually forced to make that choice, I honestly can't say what I'll do.

It's crucial I don't say anything in these interviews that will piss off the Iranian or the U.S. government, which leaves very little left to say.

I can't talk about my breakdowns in solitary confinement, I can't talk about Shane being beaten by the guards, and I definitely can't talk about how disappointed we are by our own government's inaction on our behalf. So what can I say?

"I want to ask everyone in the world who believes in our innocence to redouble their efforts." I tell CNN, ABC, VOA, MSNBC, Democracy Now!, Fox, CBS, AP, Radio Farda, BBC, Reuters, Al Jazeera, and CBS that "I'm only one-third free."

69. SARAH

Later, in a taxi, Alex gets a phone call, says something to the driver, and we're suddenly speeding off in the opposite direction. President Ahmadinejad has agreed to meet with my mother and me for a few minutes in the hotel before he returns to Iran. Cindy, Laura, and Alex will meet with one of his close aides. We'd already dismissed the possibility that he'd see us. We didn't think he'd risk looking "soft" back home right after his usual, inflammatory speech at the UN. Now we're getting rushed in to see him just hours before he leaves the country.

I lock hands with my mom and Shane's mom, Cindy, on each side of me in the back seat of the cab. Anchored between these two women, I feel like a part of Shane is physically with me. My impression is that Cindy's calm, inner strength is the backbone that holds the "Hiker campaign" together. She can make rational decisions under tremendous pressure and fight like a bulldog when she needs to.

At this moment, the contrast between her and my mom is striking. My mom has been fighting so hard for so long, she looks like she's finally ready to collapse. She recently confessed to me that she's been sick and needs surgery. Every time I look at her, my heart is filled with a rush of protective love. I want her to rest, but I know she can't bear to leave my side.

We all get out of the taxi. Cindy, Laura, my mom, and I decided to wear headscarves — I guess the idea is to show cultural sensitivity, but I think the truth is we just want the Iranian president to like us. It's the same urge I felt in prison, to please my captors, and I now realize our families feel it too. With one hand, I adjust my mom's headscarf before we go in, tightly gripping the small manila folder I'm carrying under my other arm. I stayed up late with Alex compiling a thick packet of "evidence" that Ahmadinejad asked for in an interview he did yesterday on American TV, proving that we didn't have any intention to enter or harm Iran. I had letters from all our employers, from noteworthy figures that Iran's government respects such as Noam Chomsky, all of Shane's and my own articles printed out, along with anything I could think of to prove we weren't spies.

As we approach the building, I see a bearded Iranian man in a suit that I know I've met before in prison. I of course don't know his name or position, but I still get shivers at the sight of him. "Sarah," he exclaims, "so nice to see you in your own country."

I rush up to him. "How are Shane and Josh? When are you going to let them go?" The words spill out like fire.

"They are fine, Sarah, don't worry. Just be happy that you are with your mother now," he answers smugly.

"Will you . . . Will you please tell them you saw me? That I miss them?" I grip my mom's hand tighter.

"Yes, Sarah, I will tell them. Now, you must go — the president is waiting for you."

We pass through security and are led upstairs. Someone had the brilliant idea that we should bring him flowers. When I walk into the room, I immediately feel sick and somehow ashamed to be holding this bouquet. These aren't my friends or relatives; they are the people who held me hostage. I hastily hand the flowers to one of his guards and sit down across from the most volatile and polarizing president in Iran's history.

On the surface he is small and unassuming. He directs his attention immediately toward my mother, refusing to acknowledge or make eye contact with me. I listen politely while he congratulates my mother on my release and inquires about her health problems, which have been in the media. When he tells her she should "drink cranberry juice" to help with her gallbladder, I decide the pleasantries have gone on long enough.

"Mr. President," I say, "I've brought you the evidence you asked for." I push the folder across the table.

"What is this?" he asks, looking at me for the first time.

"You told the news two days ago that you wanted evidence that

Josh and Shane are not spies. Well, here it is. The three of us are very critical of our government's foreign policy. We are all against the wars in Iraq and Afghanistan. I even helped organize a protest against bombing Iran when I was in my twenties. Will you tell this to the judiciary?" He takes the folder and begins to rustle through it in silence.

"Do you think we are spies?" I ask.

"Sarah." He pauses. "I know about your activism. You are friends with this American man that was killed by the Israeli military."

"Tristan Anderson. He wasn't killed — he survived."

"You are good kids. I hope you will be married soon and have many children, fifteen perhaps." He smiles and directs his gaze back to my mom.

"Sir," I say, drawing him back, "how am I supposed to get married when my fiancé is in your prison?"

"Sarah, don't worry," he says, "you are home now, in your country. We're not going to take you again." The president turns to smile at the man sitting next to him, as if sharing a joke, but instead of returning the smile, the president's aide looks at him sternly and whispers in his ear.

"I mean," the president corrects himself, "we're not going to take you. You are free now." He pauses, scanning the room, seeming to gather his thoughts. "I will give my recommendation to the judiciary for the release of your friends," he continues. "Sarah, maybe you can help improve the relationship between Iran and the U.S. This might be a good job for you."

The president stands and bows slightly with his hand over his heart. As always, he is the picture of humility. I'm stunned by how sloppy the president is with his speech. "Take" sounds a lot like "kidnap" to me. What is our incarceration if not an "institutional kidnapping"? They knew from the first day that we weren't spies and that we did not even willingly cross the border, yet they decided to hold us anyway.

70. SHANE

There's a new man on our hall. They put him here a couple weeks ago, not long after Sarah left. I've never seen him, only heard him, and he barely sounds human. When he walks, he makes a whiny, pulsating sound that repeats as if it were coming from a motor. It has high notes mixed with a deep guttural groan and shallow but heavy breathing. His feet drag when he walks and if a guard speaks, his drone becomes more rapid and nervous. I have never heard him use words. At first, when guards came to his cell to give him food at mealtime, he would scream in terror, similar to the way monkeys scream in cages. Now, he only screams when they try to take him out to *hava khori*.

Sometimes he flings himself against his door. The sound it makes is hard and shrill, more like a head against metal than a shoulder. He does this until the guards come, but when they do, he screams again. When I hear him, I picture him naked in a corner with his arms curled around himself, his body turned toward the wall.

When he walks down the hall, the guards kick him and laugh like boys tormenting a cowering dog. It seems that something about him, about how pathetic he is, makes them angry. I think they hate him because he is making plain to them what it is they are party to, even if they themselves aren't the torturers. Or maybe they see him as a traitor, someone who diminished the stature of humankind by letting himself deteriorate in such a way. Maybe his imbecility shames them and their shame emerges as anger.

He makes me angry too. I hate it when he screams. I hate hearing his drone. When Josh and I used to hear someone crying out in desperation, we would put our books down and abide in awkward silence, giving him at least the decency of quiet empathy. Now, we just keep reading. At least I try to, but I can't help being distracted.

Something has been quickly changing in us since Sarah left. She kept us gentle, but now we are hardening. We speak to each other less, and an amorphous tension lingers in the air between us. Sometimes we focus this tension on each other. Other times, we target something outside our cell. Now, as this man groans, a part of me wants to tell him to shut up. He takes me out of my world of books. I rarely leave that world anymore except for food and exercise. I don't want to leave it to hear someone's torment. I want to be far away from that. He sends me into a cycle of self-loathing because I hate that I just want to close out his agony.

One day, a guard tries to take him out of his cell, but he only screams. He won't go. A half-hour later, a group of them comes down the hall. I hear metal clinking against itself as they walk, like chains or shackles. I look over at Josh. He hears it too. Seconds later, we hear the sound of a shower. He is screaming in pure panic. A guard yells angrily, almost in a frenzy. Josh and I jump up and pound on the door. We pound and pound and pound. This pounding makes me feel alive again, like running hot water over a frozen hand. The screaming stops. A guard comes to the door with rage in his eyes and asks what we want.

"In chee-eh, Guantánamo?" I say. Nothing angers them more than comparing Evin to the U.S. prison at Guantánamo Bay.

He looks at us with surprise. "What do *you* care?" his eyes say. *"Karetun neest!"* he snaps. *None of your business!* He marches off. The water stops. They take the man out of his cell and put him in the small open-air cell in our hall. It is dark and cold outside. They leave him there. He rams against the door over and over again. Eventually the ramming stops. When it does, I sleep.

71. SARAH

I'm in DC. We have half a dozen meetings lined up with officials at various embassies, the State Department, and the White House. I get a call from Salem in Oman and he tells me that he has just gotten off the phone with Ahmadinejad. "He's very happy, Sarah. He likes the media you've done. He says it is helping him. He wants to free Shane and Josh — he just needs an excuse to do it, something to calm down the judiciary."

"What, Salem?" I ask. "What specifically should we be asking for?"

"Anything will do, any return gesture from the United States to Iran. Iran has already given the U.S. something by releasing you."

I tell him we have a meeting with Secretary Clinton in a few hours.

"Very good," he says, "we need engagement. The internal situation in Iran is very bad — the president is getting all kinds of criticism for letting you go. A member of parliament recently said that releasing you was a gift to 'Quran burners.'" Salem's referring to a recent scandal that erupted when a Christian pastor in the United States threatened to publicly set fire to two hundred Qurans. "The president can't put his neck out again without something to show — maybe your Secretary Clinton can send her husband to Iran, or let a few Iranian students out of prison who overstayed their visas. Anything will work."

"Okay, Salem, but the truth is," I say, thinking about the upcoming midterm elections, "Obama's position doesn't look much better. He can't afford to look weak on Iran."

Walking into the State Department, I feel truly nervous for the first time since my release. Am I thawing out, I wonder, starting to feel more? The truth is I feel deeply conflicted about Hillary Clinton. Josh's mom, Laura, tells me Clinton has been one of our biggest advocates. She gave our moms an award on Mother's Day and she's made more public statements about us than any other high-level official. "Okay," I reply, "but what has she done to help get us out?"

I can't shake the belief that statements from Clinton only make our situation worse. After all, she's one of the most demonized U.S. officials on Iranian TV, particularly since she publicly stated that the United States would "obliterate Iran" if it ever attacked Israel. IRINN and other state-run news channels in Iran usually show pictures of her with her face contorted in anger, her eyes vacant, as if she were the very face of "the Great Satan." A part of me has begun to believe the Iranian government has been taking out its anger with her directly on us.

The woman who greets me in front of her office, with her round cherub cheeks, does not bring any of the sinister images to mind that haunted me in my solitary cell. We shake hands and she leads Cindy, Laura, Nora, Alex, and me into one of the State Department's many lavish reception rooms.

I was planning on being stern with Clinton, but somehow she disarms me with a combination of sweetness and authority that immediately sends me into internal conflict. Am I wrong to blame her? Is she really at fault?

Clinton reaches out and gently touches my knee. "Sarah, I want you to know how much I admire your bravery," she says. *Don't cry,* I tell myself, *goddammit!* "You are a very strong person."

"Thank you, Madame Secretary," I manage to say evenly.

"How are you?" she asks.

"Not good. I have no way of knowing when I'll see them again, or if they're in more danger because I'm not there. The guards and interrogators were more civil with us, I believe, because one of us was a woman."

"That makes sense," she replies with a nod.

"Madame Secretary," I say, "I know we don't have much time. I want to be direct."

"Sarah, feel free to say anything."

"The Iranian government considers releasing me to be a gesture of goodwill. If our government doesn't give anything back, they'll take Shane and Josh to trial. I don't have to tell you that espionage carries the death penalty in Iran. Their lives are at risk."

"Yes, I know, but we don't believe Iran will ever do that. The consequences would be too great."

"How can we be sure of that? The Iranian government is totally unpredictable. We want you to consider releasing this woman, Shahrazad. Or the ex–Iranian ambassador to Jordan, Nosratollah Tajik, who's been under house arrest in London since 2006. We've heard directly from representatives at the British embassy that the British government feels no need to extradite him — they'll let him go if the U.S. drops the request. Both of these cases are low-profile and neither committed a serious crime, if any at all. In Shahrazad's case, it was her husband who was the mastermind, not her. Tajik is a sick, old man. These people are not dangerous."

"Sarah, I understand the situation and I appreciate its gravity. We're looking at this from every angle. I can't tell you what we will do at this point, but I can promise you it's being treated as a top priority. We are going to do our very best to come up with something that will work."

As we get up to leave, I realize that, though my criticism about her approach to foreign policy hasn't changed, there's something about Hillary Clinton that I can't help but like. At least she heard me out, and my gut tells me she's being straight with us.

Outside, we pile into two separate taxis. "Take us to the White House, please," Josh's mom, Laura, says to the driver.

The office they lead us to is filled with dark, heavy wood furniture and oil paintings. We've been sitting here less than ten minutes when the door opens and we're told the president will meet with us. This is a complete surprise — we were told we'd meet with Dennis Ross, one of Obama's chief advisors on Iran. Instead, they lead us down the hall, onto an elevator, up two floors, then up a small spiral staircase into the Oval Office.

The president gives me a warm hug. We sit down and begin to exchange pleasantries, but I can feel the minutes ticking away. This meeting means everything, and there's something I have to say.

"Mr. President," I say, "after we were captured, in the early terrifying days, my one consolation was the fact that you were in office. As unlucky as we were, we knew we were at least lucky on that count."

I'm sort of bullshitting — but a part of me means it. I definitely felt hopeful that Obama could sweet-talk us out of prison in the early weeks of our detention, but after a year in prison, I wasn't so sure. The fact that the Iranian government freed the 1979 hostages to President Reagan is often cited to support the claim that Iranian leaders have historically favored Republican presidents over Democrats. Others argue that the Iranian government is simply good at playing one party against the other — citing the fact that the Islamic Republic waited until just hours after Carter was out of office to release the hostages in 1979. This calculated move ended the hostage crisis with a bang, and showed the world that Iran was pretty powerful — maybe even powerful enough to influence the outcome of a U.S. election.

Even if Republicans have at times gotten results from a more aggressive approach to Iran, at what cost? Diplomacy between our countries definitely didn't improve under Reagan or the Bushes. Iran's nuclear program has continued to develop and today we're just as close, maybe closer than ever, to all-out war. The Obama administration has been far from soft on Iran — it has slammed them with the toughest sanctions in Iran's history, thereby hurting the Iranian people and arguably leaving their government unscathed. Perhaps an even more aggressive stance could have gotten Josh, Shane, and me out of prison sooner, but it also could have done the opposite. Either way, any strategy that means inching closer to war would never be worth it.

The president's face looks slightly dismayed, almost as if he's read my thoughts and is privately agreeing with my logic. "Thank you, Sarah," he says. "I can tell you this much. Your case has been a priority in this office. We discuss it daily."

"Sir, I'm in close contact with the Omani envoy who negotiated my release, Salem al-Ismaily. He said that Ahmadinejad is ready to do for Shane and Josh what he did for me — he just needs a gesture of goodwill from you. In Iran, I met with the daughters of a woman who's being imprisoned here in the U.S. Her name is Shahrazad." I realize I don't have the letter they gave me — I left it with Salem. "It looks like she's doing time for assisting her husband, who tried to smuggle night-vision goggles into Iran. If she could just be shown leniency, released early, she's already done three years and deserves to go home to her family. This kind of thing could be done quietly, and it would work."

"I'm sorry, I can't discuss the specifics with you of what I will or won't do," the president says, standing up and buttoning his jacket. "If you'll excuse me, I have another meeting. Just know we will do absolutely everything we can until Shane and Josh are home."

The president puts out his hand for us to walk out first, and Cindy, Laura, and Alex stand up. When we reach the door, he stops and shakes each of our hands a second time.

"Can I ask you a question?" Obama asks, cocking his head and giving me a warm smile. I get the feeling he's busy but doesn't want to end our meeting too abruptly. "What did you miss the most? Was there a certain food you really wanted, something like that?"

"Wide open spaces," I say, without hesitation. "Like the sky and the ocean. I always dreamed about the ocean." The president nods.

As I walk out of the White House, I think about what I came out of a few weeks ago. My world was so small. The shape of the plaster on the wall of my cell — in one spot it looked like a woman lying prone, in another like asparagus — mattered to me. The sound that another prisoner's footsteps made as she passed my cell had an emotional impact. All those details became the stars in my sky, the only meaningful backdrop to my life. Now, my world is bigger than ever. I've gone from seeing the world through the eye of a needle to standing on the steps of the White House.

I have to be careful, I think. I can't let all this attention affect me.

In less than a week I've met with the presidents of Iran and the United States, the secretary of state, countless diplomats, foreign ministers, and gone on *Oprah*. I have to tease out what's meaningful here and what isn't. I feel a pang of longing for Shane and Josh, and immediately I'm grateful for that feeling; my love for them will continue to orient me, to guide me on the right path.

72. SARAH

I'm about to step into a hot shower in my hotel room when the phone rings. "There's been a complication with your mom's surgery," a friend tells me. "It only happens in two percent of gallbladder removals. Her pancreas was triggered and now it's blown up like a balloon. They say she'll be okay in a week or two, but in the meantime there's not much they can do and the pain is severe."

My mom never mentioned she was sick in any of her letters to me in prison. She put off her surgery until I got home — partly out of fear that something might happen to her while under anesthesia. In the last eight months she's had to go to the emergency room several times after being immobilized by attacks. I witnessed one in the airport on the way to Chicago for *Oprah*. My mom was in the restroom stall for twenty minutes, gasping and crying as she tried to breathe through the pain.

I rush to the airport. I step up to the porter and hand him my passport. He looks at me strangely. I feel ashamed to be clear across the country right now. Even though the doctor had said she was in the clear, I shouldn't have left her.

"That will be twenty-five dollars for your bag check," the porter says in a slow, Jamaican drawl.

"What?" I ask, giving him an irritated look.

"To check your bag you have to pay twenty-five dollars," he repeats, smiling.

"Since when?" I ask indignantly.

"For a long time now." He pauses and his smile gets even bigger. "I know you. You're Sarah — you've been in Iran."

His sweetness disarms me. "Sorry," I say, "I guess things really have changed." I take out my wallet and hand him my credit card.

"Don't worry," the porter says, reaching out to take my card, "I know your boyfriend will be back soon. You're famous!"

I smile and thank him. I've been out only three weeks, but this is already the fourteenth time I've boarded a plane. I went to Georgia to see my sister and her family, to Oakland to be reunited with my friends, to Minnesota so I could wrap my arms around Shane's sisters and his sweet, teddy bear of a dad, to L.A. to be filmed for a documentary, back and forth to DC and New York several times — and now I'm on my way back to Oakland.

This time must feel like an eternity for Shane and Josh. I wonder what books they're reading now. When I left, Josh was about to tackle *Ulysses.* I suddenly get a pang of sadness when I realize I didn't ask Shane what he was going to read next the last time I saw him. If I knew what book Shane was reading, I could imagine him sitting with his back to the cell wall, the book perched on his bent knees, slowly and carefully turning each page, coaxing minutes into hours, hours into days, days into weeks. How long before they let me hear his voice?

My friend Bessa is waiting for me at the airport in Oakland. She's one of the few people I know who can roll with me in my hypervigilant and wildly unpredictable emotional state. I get into her car, and Bessa listens without comment as I vent for ten minutes. Then she smiles and asks if I've eaten. She's a nurse, and she tells me that in most cases like my mom's, the pancreas will eventually return to normal.

"Most cases?" I ask, looking at her intensely, almost accusingly.

"It's gonna be okay," she says. "Just breathe."

I rush into my mom's hospital room and find her doubled over in pain, small and weak and barely able to talk. For the first time since I was released into her arms, I feel the incredible satisfaction of knowing that I'm exactly where I need to be.

Over the next few weeks, a constant stream of friends stops by the hospital with food, money, and clothes for me and flowers for my mom. This tight-knit community has been holding my mom up for more than a year, and I know they'll do the same for me as long as this takes. The nurses set up a cot for me in the room and my friends take shifts running my errands, taking dictation, and filling me in on what I've missed. Prison has reduced my skill at dealing with the basic details of life — like remembering to sleep and exercise. I can't imagine how impossible this would be without all these people taking care of me.

Sitting next to my mom, I arrange and rearrange her pillows in a futile attempt to find a position that lessens her pain. I'm amazed at the incredible work that's been done on our behalf. Our families and friends have been able to shape the narrative around our story. Though the title "American hikers" grates on me every time I hear it, I have to admit it was a media-savvy compromise, considering that journalists and activists were both volatile categories in post–Green Movement Iran. It also made us famous. When you type "hiker" into Google, we're the first thing that comes up. Very few prisoners in history have had that kind of branding.

In the hospital — with my BlackBerry pressed to my ear and my other hand constantly jotting down notes — I learn that Nicole, Shane's sister, has a room in her house where she sorts Free the Hikers mail, copying every letter of support, sending them to Evin Prison, and filing away the originals for when we return. She also forwards books to the State Department to be delivered to the Swiss embassy in Tehran by diplomatic pouch. Shane's father, Al, is a mechanic who rarely touched a computer before our capture. Now he boasts about having accumulated six thousand e-mails over the last year. Alex, Josh's brother, has become fluent in diplomatic double-speak, and Shannon, Shane's sister, regularly talks with an anonymous caller who offered to pay our bail. Our families have become well-oiled machines.

Still, trying to make strategic decisions on conference calls is like herding a gang of angry, traumatized cats. Josh's parents are not at all afraid to express their distaste and discomfort on the calls — which often rubs Shane's family of quiet, stalwart Minnesotans the wrong way. My unruly family members constitute the wildcards — no one ever knows what they're going to do or say. Our politics are as different as our personalities, and the calls are usually a mixture of tears, awkward silences, rants, and interminable monologues. At times it

feels like ten egos competing for space, and I have to constantly remind myself that we're all doing our best and we're doing this for
love.

I've spent the better part of a week combing through e-mails, dig-
ging up old pictures, and collecting various testimonies to support
the veracity of our story. The result is a thirty-five-page document
written for our lawyer, Masoud Shafii, that he can use if and when
Shane and Josh are brought to court. I've tried to address any suspi-
cions that came up during our interrogation — trips to Israel and Palestine, my employment at a language school in Syria, and, of course,
the details of our visit to Iraqi Kurdistan. Sitting in an empty hospital room the nurses let me borrow so as not to wake my Mom from a
rare nap, I get on a conference call with our families so I can answer
any questions they might have about the contents of my research.

"I read the document," Jacob, Josh's dad, begins in a serious voice.
"What is this about 'occupied Palestine'? Are you an idiot?"

"Did Jacob just call me an idiot?" I ask coldly to no one in particular.

"Jacob, calm down and ask your question," Cindy breaks in.

"Jews founded Tel Aviv in 1909 — it was sand dunes before that.
How can you say it's *occupied?*"

"Dad, calm down," Alex says. "Listen to Sarah before you assume
anything."

"Jacob, I'm not about to have an argument with you about Israel
and Palestine. This document was written for our lawyer, to be used
in Iran's courts," I retort, starting to lose my cool. "The Iranian government refers to *all* of Israel as 'occupied territory.' This document
is not for the media — it's for the Revolutionary Court!"

We all take turns losing our cool on these calls — and I don't think
there's one of us who hasn't said things we regret. None of us were
prepared to have our lives taken over by an international crisis. Yet,
despite the volatile, high-stress nature of our campaign, incredible
things have been accomplished. Over the course of a year and a half,
our friends and families have managed to turn an avalanche of criticism, blame, and misinformation into a campaign that boasts thirty
thousand Facebook followers and high-level support from countries
such as Brazil, Turkey, Oman, and Senegal — as well as figures like

Noam Chomsky, Muhammad Ali, Ban Ki-moon, and Desmond Tutu. On the one-year mark of our detention, with vigils being held for us in thirty-five countries around the world, President Obama made a statement in which he referred to Shane, Josh, and me as the "best of the American spirit," a clear indication of the positive shift in public opinion that had taken place during that time.

As I sit at my mom's bedside, combing her long, silver hair, I'm overwhelmed by how it feels to know that our lives mean this much to so many people. Shane's and Josh's families are fighting like lions, just like my mom did for fourteen months. Now that her fight is finally done, she has collapsed like a tent in a storm. This is the same woman who worked the night shift at a local restaurant to put herself through nursing school and raised me without help from anyone. I take her hand and press it to my lips, promising myself she'll never have to fight like that again.

If my mom and the families hadn't managed to draw so much attention to my health concerns and solitary confinement, I wouldn't be here right now. We have to figure out how to do the same for Josh and Shane. It should be easier to get them out now that Iran's admitted I'm not a spy — but they still need to save face. Shane has stomach problems, but Josh doesn't — we can't play up a health issue and risk having only one of them released. The only way is for the United States to give something, but will it be possible to get these two governments to agree on anything?

73. SHANE

The interrogators came and gave us pictures. There is Sarah standing in the woods with my mom and sisters; Sarah with Josh's brother, Alex, in a parking lot; Sarah in a beautiful, busty orange dress at some kind of restaurant. She looks so strong and healthy. There is freedom on her skin and conviction in her eyes. Her smile is stunning. I don't think she ever smiled that way in here.

These images bring a flicker of life and happiness to me — a spark in my chest that makes my breath lighter for a while. It opens my imagination and makes conversation flow between Josh and me for hours. We walk briskly at *hava khori* and try to picture her life. Is she living in Minnesota with my family or is she in California? Is she re-cording her songs? Is she on the news a lot? Does she go to the beach? Has she met with Obama? She must be telling our families every de-tail about prison. They must call her whenever they have a question about what life is like for us. Having that link between us and them must put them so much more at ease. I wonder if she talks to me at night before falling asleep in the same way that I whisper under my breath to her. I wonder if she rolls up a blanket and cuddles up to it like I do. What does she have that connects her to us? I wonder if she feels alone out there, where no one understands what she's been through. I wonder if she talks to anyone about it. I wonder if she rests. How is she surviving? People need money out there, jobs.

At least she was not damaged by her time in here. Her pic-tures — her eyes — reassure me of at least that.

I ask the guards for tape and stick Sarah's pictures to the wall by my bed. I stare into her eyes before I fall asleep and gaze at her when I wake. But as the days pass, she becomes more and more two-dimen-sional again. My feelings become harder to reach beneath a thicken-ing layer of numbness. I can look and I can appreciate and I can know that there is love and yearning somewhere inside me, but I can't ac-cess it. And I don't try to. It is better like this, like the way people freeze themselves in science fiction movies to come back to life later when the world has improved. I can live with the regular sadness that I wake to every day. I don't need more emotion than that.

74. SARAH

In the front seat, Alex is rapping Dead Prez: "The White House is the rock house." He leans toward the cabdriver, moving his hands to mark the beats. "You're a musician?" the driver asks, prompting Alex to launch into one of his long Dylan renditions, reminding me of Josh in a way that both stings and comforts. Laura's talking nonstop to no one in particular and Cindy is grumbling under her breath. It's already mid-December and it's my fourth trip to DC. Everyone else says they've lost count, but it's probably close to their thirteenth. I know in a sense it's a rare privilege to get this much face time with the

U.S. government, but I'm starting to wonder just how much face time it will take to get actual results. We're here to ask the United States again to extend some kind of good gesture toward Iran and this time we have ideas straight from Iran.

"You're going to like Dennis Ross," Laura tells me. "He's been wonderful."

Laura Fattal is a proud mother. She's somehow managed to keep a sunny disposition and optimistic outlook on life even with her younger son being held hostage. I'm amazed by how resilient she is, how skillfully she lets the stresses of the campaign roll off her shoulders. Still, from what I've read about Dennis Ross, I can't imagine I'm going to find him to be "wonderful."

The more I read online about him, the less I understand why he, of all people, has been appointed to our case. Ross has a long history of siding with war hawks — and Israel — on almost every issue. In the 1980s he cofounded the Washington Institute for Near East Policy (WINEP), a think tank widely viewed as an offshoot of the American Israel Public Affairs Committee (AIPAC), the group at the heart of America's Israel lobby. This morning I found an article online about the first paper Ross published with WINEP, in which he argues for the appointment of a "non-Arabist special Middle East envoy" who won't "feel guilty about our relationship with Israel." Just a few years later President Clinton appointed Ross as his special Middle East coordinator, a position from which Ross could happily take his own advice.

Ross has held top positions ever since — playing a key role in Israel-Palestine negotiations, supporting war in Iraq — and he has now been appointed President Obama's advisor on Iran. "It means flying to Tehran by the connecting flight vis-à-vis Tel Aviv," said Sadegh Karrazi, one of Iran's top negotiators, when Obama brought Ross from the State Department to the National Security Council last year. He's one of the White House's most vocal proponents of crippling sanctions against Iran — guiding the Obama administration down the path of dramatically increasing sanctions before less aggressive means of diplomacy even had a chance. Walking up the White House steps, I struggle to put my feelings aside. Though I'm clearly at odds with Ross's politics, everyone — from Salem to Secretary Clinton — has been stressing the power this man has over our situation. "The buck stops with Ross," Salem told me while we were strolling on the beach in Oman. *Just feel him out*, I tell myself. *Whatever you do, make a good impression.*

We're escorted into the room, and Ross motions for us to sit down across from him and two other representatives from the National Security Council — Puneet Talwar, the senior director for the region, and Mustafa M. Popal, the director for Iran. The room is so staid and formal, a part of me expects one of them to take out a cigar and begin puffing away. I look at Ross's weary, beleaguered face, searching for clues. My initial impression is that he hates this part of his job — meeting with bereaved and traumatized families — that he doesn't want to be in this room.

"Dr. Ismaily's heard back from Ahmadinejad and they have a proposal," I begin. "They understand that you won't release any Iranian prisoners, so they have come up with something else. Salem says a letter, from President Obama to his Iranian counterpart, is all it will take to get Josh and Shane out. It doesn't have to contain any concrete promises, not even a hint of the possibility of engagement, just a few innocuous platitudes about peace and cooperation between nations. Salem has taken the liberty of drafting that letter. I have a copy here," I say, pushing the letter across the table.

Ross and Puneet Talwar exchange a heavy glance. It's not the reaction I expected. "We think we can do better," Ross says after a long pause, reaching over to take the letter. "We're going to designate Jundullah, the Iranian dissident group, a Foreign Terrorist Organization. It's something this administration has been meaning to do — something we honestly should have done a long time ago. Hopefully it will be just the good gesture we need to get them to make up their minds and release Shane and Josh."

"Putting Jundullah on the list sounds like the right thing to do, but how do we know it will work?" Alex asks.

"Is this something the Iranian government specifically asked for?" I add cautiously. Jundullah is the organization responsible for a suicide bombing in 2009 that killed fifteen Revolutionary Guards in south-east Iran. It's grossly hypocritical that, two years later, the United States still hasn't put them on the list. It's also an example of how politics trump everything else in the United States' War on Terror.

"No, but we'll let Salem know, of course. Anyway, it will be public soon enough," Puneet says.

"We're still asking you to write the letter," Cindy says, then adds, "We trust Salem when he says it will work."

I understand a letter is more than a letter, and we have to expect Iran would go public with it, but I'm still shocked by Ross's resistance. They should be willing to do something a little outside their comfort zone by now. After all, the safety of two American citizens is at stake.

"What about the positive statements recently made by Iran's chief of human rights, Mohammad-Javad Larijani?" Laura cuts in, changing the subject. "Isn't that a good sign?"

"Yes, we hope so," Ross answers, "but you can never tell from these statements if there's actually any consensus behind him."

During a recent visit to New York, Mohammad Javad Larijani took credit for my release in an interview on NBC, stating that he was able to convince the judiciary that I was "incapable of espionage." He also blamed Iran's Ministry of Intelligence for "overreacting" to our case and said he hoped this issue could be settled outside of court. "Let us assume that these people are innocent, that they were really hikers," he added.

Larijani's two brothers control Iran's judiciary and parliament. Some experts speculate that the three are our biggest obstacles, because the Larijani family members are the president's rivals (Ali Larijani lost to Ahmadinejad in 2005). Salem has told me that Ahmadinejad's original plan was to pardon all three of us last September — he was even planning on publicly presenting us with gifts before flying us to New York on his personal plane — but the judiciary stepped in to block him. The last thing the Larijanis want is for Ahmadinejad to get all the "compassion points" from another release.

Outside the White House, I immediately call Salem and tell him the news. "Sarah dear," he says, "this is good news and it might work, but I must consult the Iranians first. They have to approve this before I can go back. I can't just spring it on them."

75. JOSH

The evening call to prayer blares from the loudspeakers. That's my signal to grab my little book on Buddhist wisdom and read today's quote. Shane grabs his copy and also reads silently. We agreed with Sarah to do this every evening during the call to prayer. She would buy a copy and read too. It'd be a way to connect to each other and to the world outside Section 209 in general. Tonight, as usual, the Buddhist quote encourages me to live in the present moment and to have courage to face the unknown.

The three of us used to read these quotes together at *hava khori,* but now we read separately. Shane and I have been taking more space from each other. We do this especially when small disagreements arise. There's no need to get angry and dispute where Saddam Hussein was born or what floor of the prison our cell is on or whether two books can legally have the same title.

Shane and I are slipping into an intimacy that only couples know. I wish I could tell Sarah that I understand her perspective more every day and that I think about her so much. But I've wondered if she thinks about me and not just Shane. I wonder if she spends time with my family too. I bet it has gotten harder for her to include me the way she used to when she was in prison.

For over two months we've not heard a word from Sarah besides that short TV clip at the press conference. I think Dumb Guy is trying to break us down with neglect. He's only given me a selection of my mother's letters, though she writes every day. And he hasn't given me anything from Alex in many months.

The Buddhist quotes help counteract the daily humiliations of taking orders, wearing a blindfold, and having the door slam shut after they deliver meals. I've gotten used to it all, but still, it wears on me over time.

A new guard stops by to chat, as new guards often do. He listens to our spiel: no trial, no lawyer, only five minutes on the telephone. He leaves our cell, clearly appalled at our conditions. Minutes later he returns with another guard and asks again:

"Only five minutes on the telephone?"

"Only five minutes," I say soberly in my rudimentary Farsi. "We've been here for sixteen months."

Two days later, it's the Saturday after Thanksgiving and the interrogators allow us to call home. My interrogator takes me to the pay phone and I dial my parents' landline. My mother picks up.

"*Josh!* It's you! How are you? Are you okay?"

I reassure her about my health and safety. I tell her I'm now reading *Moby Dick,* and she says she'll get herself a copy immediately.

"Wait a minute." I stop myself. "Is anyone else home?"

She yells, "Alex! Jacob!"

She never calls my dad Jacob. She always calls him his Hebrew name, Yaakov, but she's being cautious, downplaying his heritage while on the phone with Iran.

They pick up and we all gush for a moment.

Alex speaks up. "We're doing *everything* we can. How are you?"

"Alex, 'My heart is not weary. It's light and it's free,'" I reply.

He immediately fills in the rest of Bob Dylan's verse: "'I've got nothing but affection for all those who've sailed with me.'"

It's the first time I've heard Alex's voice. He didn't used to sound so similar to my father — who chimes in: "Josh, we're going to get you home somehow."

"*Baruch Hashem,*" I respond into the receiver, using one of the few Hebrew phrases I know. *God willing.* I want to signal to them that I'm not scared of the Iranians' knowing I'm Jewish. That they'll do what they want with me, but I'm not going to hide my heritage. Using Hebrew is also a message to my father; it's another way of telling him that I love him.

Alex chimes in again. He always knows what I want. He reads me messages from my friends. Jenny sent me another message. She still thinks of me. Alex then tells me that he and Sarah may go to my ten-year high school reunion tonight.

"Wait a minute!" I say. "Sarah's there too? Put her on!"

She's visiting my parents' house! She might go to my high school reunion! And I was worried she wouldn't engage with my family and would de-emphasize my plight.

Sarah doesn't have a chance to say much. But I can feel her and my whole family — unwavering and supportive. Dumb Guy hoped to break me with neglect, but he lost.

76. SARAH

For the last two weeks I've heard very little from Salem. Every time I text him, I get a reply that he's "in Romania" or "in Geneva." "I promise to call when I get back to Muscat," he texts. When I finally get him on the phone, the news is bad. Salem went to Iran, but he had to return, obviously, without Shane and Josh — or even a timeline for their release. As he predicted, when the United States put Jundullah on the Foreign Terrorist Organization list without consulting the Iranian officials in charge of our case, they took offense and refused to connect the gesture to Josh and Shane.

Salem sat in on a meeting in Tehran about Shane and Josh, and none of the authorities could agree on what should be done. He tells me one of the Iranian officials present suggested that either Josh or Shane should be released first — in an effort to increase the pressure on the United States to give something in return for the release of the last one.

"Salem," I say, "you can't leave one of them there alone. Shane and Josh would never want that." I am almost shouting. "None of us would want that, under any circumstances. Please, you have to promise me that —"

"I already refused their offer," he interjects. "If one was released, it would be impossible to get the last one out. Try to stay calm, my dear. I swear to you I will never let that happen."

"Sarah," Salem continues, "the president asked me to tell you that it's not a matter of *if* they are going to get out — it's a matter of *when*."

Some comfort. All the work we've done for the last four months has amounted to nothing. A part of me feels like I've let myself be used, agreeing to be so diplomatic and even handed in the media. Now that Ahmadinejad is back in Iran, having basked in the attention that releasing me garnered, there's no real incentive for him to stick his neck out again. The Iranian government, as usual, plays by its own rules.

I've been living out of my suitcase and have been amped up on adrenaline for four months straight. The campaign will basically shut down for the holidays. This is the perfect opportunity for me to do my own research to try to put together some of the pieces of this puzzle on my own. I call my sister and ask if she knows of a private spot where I can hole up for a few days. She does.

The next day my sister picks me up at the bus station in Athens, Georgia. We're heading to her pastor's house, which I'll have to myself for a week, when my cell phone rings. It's a DC number.

"Sarah, it's Qubad Talabani. I'm so sorry to be calling you so close to Christmas."

Qubad Talabani is the second son of the president of Iraq, Jalal Talabani. Before he was elected president in 2005, his father was the head of one of the main Kurdish political parties, the Patriotic Union of Kurdistan. Qubad is Iraqi Kurdistan's representative in the United States. That means that when we were captured on July 31, 2009, in the mountains of Iraqi Kurdistan, we were on his watch.

I first met Qubad at the Hilton in DC a few weeks ago. He told me that he was in Iraqi Kurdistan at the time we were captured. "I got a call from our security services," he explained, "and they told me that three of my people had been taken by the Iranians. I knew right away it would be bad."

"*Your* people?" I asked.

"Yes, Americans." He smiled. "You know, my peeps." We both smiled.

Qubad went on to say that thus far, every time his father raised our case, the Iranian authorities told him it was "going to take some time." Now that the one-year mark has passed, he felt he could ask his father to try again. I nodded and smiled politely but had no expectation that anything would come out of that meeting. Now, two weeks later, his tone is entirely different.

"Sarah, my father met with Ahmadinejad in Turkey and showed him your letter. The president told him that Shane and Josh will be released *to my father.*"

"So . . . " I falter, suddenly at a loss for words. "Did your father get this in writing?"

"No, he didn't, but there will be a follow-up meeting soon. We need to stay grounded, Sarah. There is no timeline yet. You never know how long they might drag it out, but this is much, much better than what we've heard in the past."

I get off the phone and immediately call my mom at my sister's place. "Mom, Qubad Talabani just called me," I say excitedly. "They've gotten some positive signs and have decided to get involved."

"Talabani!'" my mom shouts over the kitchen noise. "Sarah, that's great! They were so hesitant in the past."

"What did you say, Mimi?" I can hear my nephew Graham in the background. "Was Aunt Sarah talking to the Taliban?"

"No." My mom laughs. "Tala*bani*, not the Taliban. This is great, Sarah."

The biggest flaw in our campaign strategy up to this point has been that it depends too heavily on action from the United States. The U.S. government didn't get me out of prison, and they aren't going to get Josh and Shane out either. In my conversation with Qubad, he made it clear that his father plans to negotiate directly with Iran and cut the United States out of the conversation. An independent third party like Iraq may be exactly the kind of diplomatic pressure we need.

77. SHANE

Our wing of Section 209 is dusty and barren. It has always felt to me as though our hall, the eleventh and last corridor in the ward, isn't supposed to exist. It has the feeling of a storage area. Big air ducts run along the ceiling so low that you have to duck at one point to avoid hitting your head. It is also darker than other halls. It feels industrial. Sometimes, when the windows are open, the air becomes coated with the smell of truck exhaust. When it's bad, Josh and I lie on the ground — the fumes seem thinner at the bottom of the cell. In this hall, the cells have none of the ventilation slots with fans that suck the heat and condensation out of the rest of the ward. The cells here feel older and more dungeonlike than they do in the rest of the prison. The locks are the bolt kind that clang as they are driven home, not the quiet key locks used in other halls. The cells here are smaller too. Ours, at nine by fourteen feet, is the largest. The others are built for solitary confinement.

We've been in this cell for eight months now. In that time, it's become oddly homey. We have a bright shawl hung on the wall as a tapestry. Pictures of Sarah and family are speckled across its pale green walls. So are other pieces of artwork we've collected over the past year: an old photo of Damascus, my brother-in-law's photo of slabs of ice jutting out of Lake Superior. The wall next to the door has become largely concealed by our fridge, television, and bookshelves made out of cardboard boxes and stolen dinner trays. The cell contains only one bed frame and we use the space underneath it to store our books, clothes, and dry food. There is also a mattress on the floor, which we stack on top of the other bed when we need room to exercise.

We recently started taking potting soil from the big plants in the main hall and planting vegetables in milk cartons. Green onion shoots soak up the sun on the sill of our window, ten feet off the ground. We can reach it because we have a white plastic chair, which the interrogators gave us when I complained of back pain. The chair also allows us to unscrew our light bulbs at night. Our little three-by-five-foot bathroom, where we shower while straddling the squat toilet, has become the lab of our alcohol experiments — various bottles full of liquid test different theories of how to make it work.

No one bothers to search our cell anymore. Guards don't even come down to this hall very often — since there are toilets in the cells, they don't need to let people out like they do in the rest of the ward. I used to like this kind of isolation, but ever since Sarah left three months ago, it's been different. I feel like we are being warehoused in this dark corner of the prison. We've gotten almost no word from the outside. The interrogators have come only twice since Sarah left and they haven't brought us any letters from her — just a few short missives from our moms. The stillness is starting to grate on me.

Lately, it's been hard to ignore prisoners' conversations on the pay phones at the end of the hall. Some people are there every week, calling home. Some of them break down as soon as they hear their loved ones' voices. Most lie to their relatives, telling them that things are great, that they are happy and content and that everything will be cleared up soon. Some of them talk baby talk to their children. I can't help but feel jealous. Why can they call their families, and we can't even get letters?

It's almost Christmas, which I find one of the most depressing times of year to be in prison. Only the beginning of spring is worse. I imagine my family struggling to stifle tears as they sit around a tree at my mom's house. Or maybe they are each alone, walking forlornly on snowy streets, lost in their own private sadness and yearning. Or worse, maybe they are caught in a tightly woven story of hope where they tell themselves we will be released in time for the holiday.

It is getting harder to access them in my mind without letters. And when weeks pass with no word from anyone outside our grimy hallway, Josh and I get depressed. Sometimes, I just stare at the numbers I scratched into the heater months ago, hoping one day to call Sarah. We start to wonder out loud whether anything is happening out there. The outside world becomes hazy and we imagine reasons for a stalemate in our situation. We start to wonder if people are giving up. There can only be so long an absent person can stay alive in people's psyches before that person starts to fade. Has anyone, in a slip of the tongue, yet referred to us in the past tense?

We need letters to survive mentally. We need them to give us that little boost — that little shot of love and hope — that helps us hang on for a few more weeks. But Josh and I both know the letters won't come unless we do something. It's easier for them not to bring us letters. It's also better for them if we lose touch with the outside. Forlorn, depressed prisoners are easier to handle.

We really only have one weapon — hunger-striking. They don't want us to be harmed, because we are valuable to them. We need to hurt ourselves to make them treat us humanely.

One day, when Ehsan comes by our cell, we tell him about our plan to stop eating in one week. He looks shocked, but he seems to think it's a good idea. He has been sending messages to our interrogators for weeks now, telling them we need to see them. They have never responded to him. When we tell him about the hunger strike, he says, "They will come, I'm sure. If you stop eating, they will come. Write a letter to them. I will give it to my boss and make a copy for them right away. I promise."

In our letter, we are amicable, diplomatic. "We hope we don't have to resort to this," we say. "We want to have a relationship of trust with you. We don't want to cause problems."

They don't come. They are testing us. They know this is a larger fight than just one about letters. They know we are rebelling against the feeling of being warehoused and forgotten. They know we want to keep them aware that they can't neglect us now that Sarah is gone. They want to win this. Their first move is to act completely indifferent. We understand that — it would be foolish of them to respond to a threat before we have even done anything.

So we strike. It's been over a year since we've done this. When Sarah was here, we were always too afraid of being separated to hunger-strike. We're still afraid of that, but Josh and I prepare by deciding who'll take which books if we're sent off to solitary. We reason that whoever has to leave the cell should take the biggest, most difficult ones. If they pull us apart, we will go ten days with no food before we give in.

For three days, we lie in our beds. I drink water compulsively, keeping my belly always full. At mealtimes, we carry our water bottles to each other's beds to sit and talk. Being hungry with Josh is far easier than being hungry alone. Our first hunger strike, during our first week in prison, felt like a fight. It empowered me. But in every strike after that, I felt that I was watching the life slowly drain out of me. It seemed like I was withering away for nothing. I was alone, no one seemed to care, and it was hard to convince myself that it meant anything. With Josh here, I don't feel lost. Time slows and my body empties, but our solidarity keeps me afloat.

On the third day, Josh's interrogator comes. He takes Josh out of the cell. I don't know what's going on. I pass the minutes by pacing, trying not to think about Josh not coming back. Then the interrogator returns and takes me to the padded interrogation room. There is a note of concern in his tone. "What's wrong?" he asks. "Are you feeling okay?" Then he lays an envelope in front of me. I open it and find letters from Sarah. My hunger disappears. Words jump off the page.

> *We had no idea what a big deal we are . . . So many expensive gifts from Oman . . . Tons of media . . . preparing for Oprah . . . We never imagined one-fiftieth of what our friends were doing . . . family members feel guilt . . . I'm so emotionally exhausted . . . Alex and I cried in the car and held each other . . . I'm going to throw some petals in the Hudson for you guys . . . You're my favorite place in the world.*

I flip through the stack excitedly. Sarah's words connect me to the world in a way no others could. I understand where she is coming from. She understands where I am going. Suddenly I feel loved and supported, not empty and lost.

Then my heart starts to sink.

"This isn't everything," I say to the interrogator. "You have hardly given us anything in three months. There are only twenty letters here. There is nothing from my sisters. Hardly anything from my parents. This isn't enough."

"Not enough?" he says. "Shane, I told you you are not allowed letters from your sisters. It's against the rules." He did tell us that last summer, when he arbitrarily cut letters from our siblings.

"If you don't give me letters from my sisters, I can't start eating again. You have to give us all the letters from our families."

He takes the letters from my hand. "If you eat, you can have these letters. If you don't, then nothing. It's your choice. It doesn't matter to me." The stack of papers, my connection to Sarah, the taste of her first moments of freedom, the reassurance that she is okay, the sight of the world through her eyes, the end to this hunger infusing my entire body — all of these things dangle before me, free to take if only I will back down.

"I can't," I say to him. It takes every nerve in my body to stand and walk back to my cell, empty-handed. When I open the door, Josh looks at me, smiling, with a stack of letters in his hands. For some reason, they gave him all of his, including the ones from his brother. Is this another way to keep us off-balance, divided? As soon as I start to explain what happened, I hear the distinct, hard-soled footsteps of an interrogator approach. I assume he is coming to strip Josh of his letters.

The door opens a crack and from behind it, the interrogator's voice says, "Step outside." After all this time, this man still never shows his face. I put my blindfold on and we step around the corner. In a near whisper, he says, "If you eat today, we will bring you the rest of your letters in two days."

"From my sisters and —"

"You will get them all," he says.

"If you don't bring them, we will hunger-strike again," I say.

"That is okay. They will come."

I take the envelope of letters and go back into the cell. When the footsteps fade away, Josh and I shout, jump, and laugh. We give each other big, tight hugs. Out here in our dark, dusty hall, in the wilderness of Section 209, we feel like we've just slain Goliath.

On Christmas Eve, Dumb Guy comes with a box of books and a stack of letters. In the interrogation room, we look through the letters and books, and we are satisfied. As promised, I get letters from my sisters and more from my parents. Josh and I decided that we should ride the wave of our victory and use it as a precedent. We tell Dumb Guy that from now on, we will hunger-strike whenever thirty days pass without them bringing us *all* of our letters from family.

Dumb Guy seems unconcerned. He has something else on his mind. He leans back in his chair, smiling, practically begging us to ask him what he is hiding. We can see him clearly — unlike Josh's interrogator, Dumb Guy doesn't make us wear our blindfolds around him anymore. "Put your letters away," he says. He hands us two envelopes. Each one has a blank card inside with a Christmas tree on it. *This is it? He wants us to be grateful to him for bringing us Christmas cards?* It feels like a sick joke.

"Write a message to your family," he says. Josh and I look at each other, asking with our eyes what we should say. "Tell them your wishes," Dumb Guy says. "We will send it to them." I am skeptical. We have written letters to our families before, on our own accord, and asked our interrogators to deliver them. They told us they did, but our families received nothing.

"My family doesn't celebrate Christmas," Josh says. "We are Jewish."

"Don't Jews celebrate Christmas?" Dumb Guy asks. "Do you celebrate New Year's? You can write the card or you cannot," he says, acting hurt. "It's up to you."

We start to write. "Just a few sentences!" he says. He grabs Josh's card and looks it over. In the tiny writing we use out of habit to conserve our contraband ink in the cell, Josh has written a list of wishes: books he wants, being more connected to family and friends through letters and phone calls, freedom.

"You cannot *ask* them for things," Dumb Guy snaps. "You can wish them good health, a happy Christmas, things like that. You have lost your chance, Josh. You broke the rules. You are done." It's always like this with him and Josh. Dumb Guy always wants to string Josh along a little bit. Make things just a little harder for him in these meetings. By telling him he can't write a card, he is really just telling Josh to start begging. Josh never begs. He will ask for another chance, but that's it.

I write my message carefully, so as not to have it taken away. I tell them I miss them. I say that I am strong and healthy. I thank them for everything they've done. And I tell them we are getting letters from our siblings and Sarah "after time for our tears," our Bob Dylan code for "hunger-striking."

After trying to convince us that Josh doesn't get to send a card home, Dumb Guy relents and gives him a piece of paper to write on. Josh, being Josh, starts his new card, trying to make his Jewish family laugh at the absurdity of sending a Christmas card from Iranian prison. "Ho! Ho! Ho!" he writes.

In our cell that night, we take chicken we've saved out of the fridge. We make stuffing with old bread and chicken broth we kept in the freezer from a lunch weeks ago. We thaw to room temperature about ten green beans we've culled from various stews. We rehydrate some dried cranberries we've had for months and add orange zest. We put some apple slices, butter, sugar, cinnamon, and crumbled whole wheat cookies in a plastic bag and steep it in our thermos until it becomes hot apple pie. We talk about the letters so much that it almost feels like our loved ones are present. Josh, being Jewish, doesn't normally celebrate this day. It kind of feels like he is giving me a gift, celebrating Christmas in Iranian prison.

78. SARAH

I write them every day. Sometimes — after a sixteen-hour-day has left me limp and exhausted — the last thing I do before I let myself sleep is call up a friend and ask her if she'll take dictation. I lie down with the phone pressed to my cheek, close my eyes, and talk to Shane and Josh about my day as my friend types. No matter what, I'll always write.

Today is different though. For the first time since I left Iran I'm not just sending my thoughts into a void — never knowing if or when my letters will reach them. For once, instead of the one-sided conversation I've been struggling to keep up these last four months, I can respond to their actual words.

I just got your Christmas cards from the Swiss. I can't believe it! I called your mom and sisters right away. I read what you wrote

*to your dad over the phone and we both started crying. What you
wrote me — along with the messages I've gotten from the Swiss and
Salem — touched me to the very core. No one knows as well as I
do how hard you guys are fighting in there. Please know that we're
fighting just as hard out here.*

As I write, I can't help but think about all the hands my letters
will pass through before they get to them. The State Department, the
Swiss . . . The thought of Father Guy and Dumb Guy reading my letters makes me livid, but there's also the small hope in the back of my
mind that our interrogators will communicate what I write to their
superiors — even that it could make some sort of difference.

We were recently notified through the Swiss that Shane, Josh, and
I will be tried in about a month, February 6. Even though I'm on the
official papers from the Revolutionary Court, there's been no talk
from the Iranian government about my returning. The last proposed
trial date, November 6, was canceled — the official reason being that
Josh and Shane "didn't show up" in court. We realize this one could
be canceled as well. Anything could happen.

*This month of pretrial limbo is frustrating, but I am not let-
ting it discourage me. I hold out faith that the judge will show compassion in line with Islamic values and release you guys before the
trial. If they don't, I maintain hope that the trial will lead to you
and Josh being released.*

*The fact that I can now have confidence that you will receive
my letters makes it much easier and more joyful to write. Please
tell Josh that I can't wait to spend time with him. I have his voice
with me all the time, same as yours, telling me that I can overcome
even the largest obstacles with grace and dignity — just like we did
inside — telling me to breathe deeply, stretch, love myself.*

I address my letters to Shane — I can't imagine the interrogators
will give Josh letters from me since I'm not family — but I feel confident Josh knows I have them both equally in mind. I try to pass on as
much information as I can without risking censorship and at the same

time be honest. I know Shane and Josh will want the bad news as well as the good, so I pass it on faithfully. Still, I think about how carefully they used to speak to me in *hava khori*, storing up soothing words for our visits, holding back words that might scrape or burn. No matter how badly I want them to stay connected to what's happening out here, I don't feel comfortable burdening them with my problems.

Later in the evening, Josh's brother, Alex, stops by the apartment where I'm staying in DC. We read the cards again, laughing about Josh's writing "Ho! Ho! Ho!" These sparse, heavily censored lines have given us far more information than we've had in months. Josh is still being playful. Shane is worried about me. The two of them were desperate but also strong enough to go on hunger strike for more letters. They are still Shane and Josh.

79. JOSH

My insomnia came back after the hunger strike. I think that's because I've allowed myself to feel too strongly. The struggle of the hunger strike made me come alive. Then came the exuberance of winning and the joy of reading all the letters.

At this point, I just want a magic pill for all the pain that floods in when I allow myself to feel. I want this onrush of anxiety to go away. And I want to be able to sleep at night. I've tried everything: I don't drink tea after 2 p.m.; I exercise until exhaustion; I don't read nonfiction at night; I don't eat snacks at night; I meditate at night. Nothing consistently counteracts the insomnia. Showers sometimes help, but sometimes they awaken me. Even masturbation doesn't help, though I do it late at night after Shane falls asleep.

I'm anxious about nothing in particular. My eyes close and my body relaxes as I lie in bed. I sense sleep nearby, but there is one final layer of tension that I can't shed. I don't know where it is. I swear it is not in my body; otherwise, I'd massage it. I swear it's not in my mind; otherwise, I'd think it through. I just need to touch the bottom of the well and let the tension go.

A psychiatrist came a few weeks ago and prescribed a barbiturate for me, but a nurse cut me off when he realized I was just hoarding

them. I took the barbiturate once and slept like a rock. I hated being sedated though. I couldn't focus on anything for two days. I don't
really know why I'm hoarding them. Shane doesn't want them either.

A different nurse is looser with the pills, and he hands me Xanax
whenever I ask. As the pile grows, the prospect of a deep sleep and a
break from routine tempt me.

On New Year's Eve, Shane takes the bottle of fermenting dates
that we hoped would turn to alcohol by now out from its spot in the
bathroom. After months of experimentation, we've yet to success-
fully arrest the fermentation at alcohol; the liquid always ferments a
second time, turning to vinegar. I taste the water mixed with rotting
dates and cringe in a way that makes Shane laugh.

When night comes around, I hand Shane three of the large blue
barbiturates, and I swallow a blue one and a Xanax. We play a movie
about capoeira called *Only the Strong* and lie back in our beds. The
performers combine martial arts with music into a subversive dance.
I'd love to be able to combine playfulness and rebellion, openness
and defiance, like they do. I'd love to be able to sing at sunrise on the
beach and escape my enemy with a backflip.

I lean with my back against the wall. Woolen blankets cover my
body up to my neck. Right now, I feel neither playful nor rebel-
lious — just focused on living vicariously through these fighters/
dancers. Shane and I watch bleary-eyed, sedated in the present moment. My world is going blank. My mental vacancy is filled with the
lights and sounds of a television screen. Shane is on his bed, and I'm
on mine. Our beds rest on the prison floor and somewhere below this
concrete, the prison rests on Planet Earth. The earth spins and cycles
around the sun, and we drift off to sleep — nodding off to the year's
end.

80. SARAH

Our campaign runs on East Coast time, so when I'm in California, no
matter how early I wake up, I'm already far behind. I jump out of bed
at 7 a.m. I've only gotten a third of the way through my e-mails when
I see one from Chris Crowstaff, one of the directors of a UK-based
human rights organization called Safe World for Women. They've

been advocating for us since the beginning. They helped our moms appeal to the United Nations Special Rapporteur on Torture about my prolonged solitary confinement. Dr. Manfred Nowak of Austria responded by sending an official letter to the Iranian government on my behalf, and my mom used the word *torture* in the media for the first time. "It is widely recognized that protracted solitary confinement constitutes 'invisible' psychological torture," she said, "especially in a case like Sarah's when the detainee does not know why she is being held or what will happen to her."

I recently asked Safe World for Women if they would start a campaign for Zahra Bahrami, the Dutch-Iranian woman I used to call Pink Lady, whom I secretly talked to in prison. During insomniac nights, I sometimes search for information about Zahra and other prisoners who helped me at Evin. I find lovely pictures of her, with bleached blond hair, glamorous Cleopatra eyes, and a pink hijab, but what I read about her case is conflicting and inconclusive. Now, I'm standing in front of the bathroom mirror, brushing my teeth with my right hand and scrolling through e-mails on my phone with my left. When I open Chris's e-mail, it's as if time stops. I suddenly feel cold, exposed, and acutely aware of my bare feet on the tile floor.

"No!" I yell. The icy feeling turns into a hot, naked rage.

I'm back in cell 25 sitting on the floor with my back to the wall I share with Zahra. I sit with my knees drawn up to my chest and my eyes closed, feeling her presence, the warmth of her proximity. Somewhere in the distance I can hear someone crying. I know that voice.

My phone crashes on the linoleum as I fall to the floor, gasping and sobbing. I feel arms around me and think of my favorite guard, tough, stubborn Leila. I can hear Zahra's laugh in the background, her husky voice speaking to me through the vent. *"Sarah, I know your mother. I'm a mother too. I'm so sorry for you."*

The e-mail from Chris tells me that Zahra was hanged early that morning. Neither her family nor her lawyer was informed that her execution would take place. All we know is that agents of the Ministry of Intelligence secretly ferried her body to Semnan, a city four and a half hours outside of Tehran. They didn't tell her family until the last minute, so they wouldn't have enough time to get to the funeral or be able to see her body before she was buried. No one knows what happened to Zahra, what they might have done to her, in the last days and hours of her life.

"Don't forget I love you, Sarah," Zahra wrote in her last note before she was transferred. *"No matter where they take me, I will look for you. Remember to listen for me. I will call out your name at night."*

Zahra loved me. In a time and place where it should have been impossible, she brought joy into my life and reminded me that our joy was defiance. She never gave up. I often heard her yelling and arguing with the guards, fiercely defending her dignity in an inhuman place. She sang to me at night, Michael Jackson's "You Are Not Alone," and passed me notes in the bathroom trash. She spent the end of her life, not just concerned about her own pain, but trying to help me.

Immediately after I read Zahra's last note, out of sheer paranoia I ate it. Maybe a part of me knew that she was truly gone. Perhaps I wanted to be sure that the life of this fearless woman would forever be a part of me.

I spend the rest of the day scouring the web for more information about Zahra. I find out she was forty-six years old when she was murdered. She was the mother of two daughters and a son. She was born in Iran in 1965 and, after a difficult divorce in her early twenties, she moved to the Netherlands. There, she worked as a professional belly dancer, traveling back to Tehran frequently to visit her children.

Articles say she was picked up shortly after the Ashura protests in late December 2009, just like she told me during our first conversation through the vent between our cells. Instead of being charged with political crimes like other protesters, Zahra was convicted of smuggling drugs from the Netherlands into Iran.

Perhaps the most shocking thing I learn about Zahra's execution is how fast they did it. A death sentence in Iran can take years to carry out, and many are reversed on appeal. Zahra was executed less than a month after her sentence was announced, catching her friends, family, and even the Dutch government by surprise. Hers is the sixty-sixth hanging this month, a huge spike even in a country that has the highest per capita execution rate in the world. The Iranian government is on a killing spree, terrified that its people will be emboldened by the Arab Spring, the powerful popular uprisings across the Middle East. The Tunisian government has already fallen; the fear is that Iran may be next.

Clearly, the Revolutionary Guard is cracking down on dissidents, but what does that mean for Shane and Josh? As usual, it seems impossible to know. I find a radio interview attributed to Zahra, recorded just days before her arrest. In the interview, she speaks indignantly, testifying that she saw three young people run over and crushed to death by security trucks at the Ashura protests. She goes on to brazenly advocate violence as self-defense against government security forces. "Yes, we will get beaten, but we will beat them up too," she says. "I was one of the ones on the day of Ashura standing in the front lines yelling at people to attack! Do not run away! I was the one who stomped on one of the motorcycles of the Special Forces. And I was beaten as a result. I am proud of every bruise on my body. I am proud of my legs and arms that are now blue-black. I am proud of my torn-up vocal cords because chants of freedom came out of this throat!"

A few days after that interview, Zahra was dragged out of a car by her hair and taken to Evin Prison. She was charged, like many others, with "taking part in a banned protest," but these charges were never brought to court. A few weeks after her arrest, security forces searched Zahra's family's apartment and allegedly found stashes of cocaine and opium. She was subsequently charged with "possession of drugs," which was later changed to "trafficking of narcotics," which carries the death penalty; the charges related to her participation in the Ashura protests were never addressed.

Were the drugs planted? Would anyone bring drugs into Iran from Europe, especially at such a volatile time? The international outcry after her execution, including the Dutch government severing all ties with the Islamic Republic days after her execution, forced the Iranian government to respond. They point a finger at her former drug charges in the Netherlands, for which Zahra served two years for possession in 2003. As soon as this information from her past surfaces, outrage at her death deflates, but the Iranian government's message is still clear — Zahra's death is a warning. If this regime is bold enough to kill a protester, especially a woman and dual citizen, it will kill whomever it likes.

Zahra was held near me in Section 209, a ward reserved solely for political prisoners, not drug-related crimes. She was interrogated in the same rooms I was, and she told me herself that she was tortured, barely able to stand on the sink and talk to me through our common vent. The Iranian government denied her the legal right to an appeal, an opportunity for the Dutch government to intervene on her behalf since she was a dual citizen, and forced her to confess in a television interview, a confession she later recanted. Zahra's lawyer, Ms. Jinoos Sharif, stated that Zahra had been deceived, as is common for political cases in Iran. "The interrogators had promised her that if she admitted to the crimes, they would release her."

One day in *hava khori* I reached down to cup my hands around a small gray moth almost camouflaged on the dusty floor. Its wings began to flutter, and from the tattered folds emerged the most vibrant colors, bright pinks and deep greens in a pattern I'd never seen before, as intricate as a Persian rug. Zahra was like that for me. Like the moth's wings, she was a window into an Iran as beautiful, warm, and complex as my experience was cold, cruel, and barren.

I want to speak out for Zahra. I want to tell the world just how beautiful and strong she was. I want to tell them not to believe the lies told about her in court. But I know I can't. I can't risk angering the Iranian government with Shane and Josh still there. This must be how so many Iranians feel, inside and outside their country; if they fight back, they might be next, or worse, the government might retaliate on someone they love.

81. JOSH

Our television turns on automatically at 10:59 p.m. We programmed it for that time because the IRIB airs English news from 11:00 to 11:15 p.m. On the screen, a swastika transfigures into a blue, six-pointed Star of David. Barbed wire then chokes the star, which oozes blood. The graphic always comes on before and after a particular talk show.

I see this gruesome graphic on TV every week. I normally just ignore it. My Judaism was only an issue during interrogation. Since then, many guards go out of their way to remind me that Moses, Jesus, and Muhammad all reference the same singular God. But this time I keep stewing over that bleeding Star of David. A masochistic part of me wants to understand what those idiots are saying on their talk show. When I see such demonizing images, I'm reminded that I'm vulnerable to being singled out. I have no idea if sacrificing me is even on the negotiation table, but they have tried to separate Shane, Sarah, and me in different ways. I fear what's next. Maybe they'll let the couple reunite and leave the Jew behind.

I turn to Shane for reassurance. "They better not free just one of us."

Shane doesn't hesitate. "I told you, Josh, I'd never leave without you."

He did tell me, but that was a while ago. I need to hear it again. I change the channel and zone out to images of Ahmadinejad cutting the ribbon for the opening ceremony of another cement plant and of Khamenei condemning the conspiring forces of America and Zionism.

I've repressed this fear of being left behind for so long, I almost forgot about it. But it creeps back with the same potency it once had in solitary. I imagine the room without Shane, like the times when he visits the doctor. At first, I enjoy the privacy for several hours, but when he's gone longer, my dread of solitary confinement creeps in.

Shane's freedom would be my misery. If they released him, God knows how long they'd keep me — totally alone.

"Josh." Shane looks at me squarely and repeats, "I would never leave you."

"But if they try to separate us," I urge, "we should refuse to go."

I look at him intently. I know I'm being paranoid, but I can't help it.

"Honestly, Josh, I'll grab the bedpost. I'll fight to stay in the cell with you." I take a deep breath. Shane adds, "But you know, we won't have a choice of whether to be released or not."

I start to get angry at Shane. I'd be alone, like during the first month when he and Sarah spoke to each other through the vent and I suffered by myself. I've tried to forget that month, but I can't.

Shane's right though. I don't know why I'm getting mad at him. He has reassured me as best he possibly can. He even offered his body to get pummeled by AK for me. I know he wouldn't choose to leave me and that they won't give us a choice. They never give us choices. I look away from Shane, trying to disguise that I'm mad at him for telling me the truth.

82. SHANE

We have been preparing for our trial for two weeks, ever since we learned from our families' letters that we were going to court. Today is the day, February 6. As we walk out of Section 209, the janitor wishes us good luck. *"Inshallah azadi,"* he says. *Freedom, God willing.* Outside, three young, fit guards are waiting for us. Each of them is holstering a pistol. We know only one of them and he hasn't shown his face in Section 209 for a while. He carries himself with authority now — he must be moving up in Evin's bureaucracy. As soon as Josh sees him, he tells him he took the neck-exercise advice he gave him a while ago. Now, Josh lies on his side in bed with water bottles strapped to his head with a towel, and lifts the bottles up and down. I do it too. We've been doing it more than usual in anticipation for court. Neck muscles are the only muscles the cameras will see.

As we drive to the courthouse, the guards talk about cars — which are the best, how much they cost, and which one a friend of theirs just bought. Josh teases the red-headed guard sitting next to him, who is constantly texting on a cell phone, the screen of which is filled with Ayatollah Khamenei's half-smiling face. Josh asks him if he is chatting with his girlfriend. The guard shoots him a stern look, then goes back to texting. Josh pretends to fall asleep and lays his head on the guard's shoulder, which makes him jolt. Josh laughs, the guard cracks a smile, and the tension lightens.

There is no one outside the courthouse when we pull up. I have been expecting a throng of media, but the only person there is the lone guard who opens the gate. When we get out of the car, a guard unlocks the handcuff that has my wrist linked to Josh's. They take us inside and down the big echoing hall of the courthouse. After we enter, I see our lawyer, Masoud Shafii, sitting all alone in a row of seats against the wall. We haven't seen him since the day before Sarah was released. He is wearing a suit and has a briefcase on his lap, but he is slouched a little, looking at the floor like the kid in school everyone picks on. As soon as he sees us, he suddenly becomes animated. He stands, his chest inflates, and he takes a bold step forward, reaching to shake our hands. Immediately, one of the guards gets between Shafii and us, and orders him to step back.

"These are my clients!" Shafii protests in Farsi, looking outraged. His indignation seems slightly put on. He is going through the motions, acting as though justice were a thing he actually expected, even though it isn't.

Our guards brush him aside and sweep us into a medium-sized, brightly lit courtroom. The few people sitting in the gallery, separated from one another at irregular intervals, don't look at us as we enter. No one is speaking. Everyone faces the empty bench blankly. I know that this is probably the most decisive moment in our case, but when we sit, it feels like we are anonymous spectators awaiting some kind of mundane spectacle, like a man coming to announce the scores on our driving exams or read us a report on trends in the real estate market.

A translator sits with us. He is a friendly forty-something man, sharply dressed in a suit jacket and a burgundy turtleneck. He has a quiet demeanor, sharpened by his round spectacles and trimmed goatee. "I have seen you on TV," he says with a bashful smile. "I also saw Sarah. Is it true that you were just hiking in Iraq?"

"Yes," I say with a huff. "Where is our lawyer?" The defense table is empty. Are they going to let him in? Did he just come to the courthouse like a beggar? Our interrogators promised us we would be able to meet with him before our trial, but he isn't even here. There is a man behind the prosecutor's podium to our left — short, gray-haired, bespectacled, and wearing a suit jacket and a maroon scarf hanging over his neck. He appears to be sifting through notes, diligently looking away from us.

The only person in the room who seems alive is a young Press TV photographer I remember from our moms' visit — who unsuccessfully tried to get us to smile then by saying we were more famous than President Bush. He is buzzing around snapping photos, beaming like usual. Whenever Josh or I make an expression that suggests frustration or distress, the shutter snaps in a flurry. Two videographers have their cameras trained on us.

Eventually, our lawyer enters. He walks directly to the defense table ten feet to our right, sits down, and arranges his papers. Why isn't he looking at us? I try to grab his attention. "Mr. Shafii," I say in a loud whispering voice. "Can we speak?" I move to get up, but he motions for me to sit.

"Wait," he says, and looks away.

We have never been told anything about our trial by anyone in Iran. If it weren't for our families' letters, we wouldn't have known it was coming. We don't know how the Iranian judicial system works, or anything about Iranian law. And our lawyer seems afraid to make eye contact.

The judge enters. Josh and I make to rise, but stop ourselves once we see that no one else does. Something about this man's presence gives me a slight sense of relief. I've read accounts of the trials held against members of the shah's government in the early days of the revolution. They left me with images of courtrooms presided over by spindly, severe mullahs who sentence one person after another to death before jeering crowds. This judge, however, is wearing a white reticulated turtleneck sweater and gives off an aura of calm. His jowls have a studious droopiness to them and they are lightly coated with whiskers. His hair is short, black, and neatly combed. He sits down his heavy frame and peers evenly over his papers. Small portraits of Ayatollahs Khomeini and Khamenei flank him on the wall behind. A poster hanging on the front of his bench reads "Branch 15 of the Islamic Revolutionary Court."

A man enters, sits at the judge's left, and opens the Quran. He reads from Surat Al-Ma'ida — a chapter that stresses the importance of being equitable in judgment — in a beautiful rolling elocution.

When he closes the book, the sparse crowd chants in unison, *"Allahooma salle alaa Muhammad va aale Muhammad,"* calling on God to bless Muhammad and his progeny.

The judge, in monotone, reads us our charges of espionage and illegal entry and warns us to be respectful of the court. "Is Sarah Shourd present?" he asks our lawyer, as if he doesn't know. "Since she is not present, she will be tried in absentia," he says. He then instructs the prosecutor to give his opening speech.

Suddenly this short, quiet man breathes fire. He shouts and thrusts his finger into the air, but I don't know what he is saying. Our lawyer jumps up and starts complaining. The interpreter is jotting into his notebook. What the hell is going on? The interpreter leans toward Josh and me and says, "The prosecutor says on the twenty-eighth of August, you three went to Baghdad to meet with the CIA. Now your lawyer is objecting."

"Why are you lying?" Shafii is saying to the prosecutor. "They were in prison on the twenty-eighth of August!"

"I didn't say August, I said *août*," the prosecutor counters.

"*Août is* August!" Shafii spits. The judge tells him to sit down.

Josh and I look at each other. *They are saying we went to Bagh-dad?* Josh's face doesn't show fear or anger, but a look of understand-ing, like he is acknowledging that we now know which of our make-believe court scenarios have turned out to be true. A long-standing question has finally been settled — they aren't going to bother spin-ning the events of our real lives into something that can make us look like spies. They are simply going to rattle off their own story. I'm not nervous, oddly, and I don't think Josh is either. This fiction is the air we have been breathing for a year and a half. And no matter what they say now, I still believe this trial is going to do us good. Josh and I have always believed we'd need to go to trial before they would ever release us. We just need to watch them play out their show and try not to piss off the judge.

The prosecutor continues. He aims his speech not at the judge, the one he is ostensibly presenting his argument to, but at us. He looks in our direction the entire time, delivering his sermon like a fire-and-brimstone preacher behind a pulpit. "After receiving their mission from the CIA at the U.S. embassy in Baghdad, they went on a mission to the north of Iraq to prepare a report from Iran's border-line regions," he shouts. "After their arrest by the Iranian border po-lice, the head of Iraq's Kurdish police confirmed their cooperation with the CIA."

At this point Shafii jumps out of his seat again and interjects. "How do you have this information on the Kurdish police?" he asks sharply. "Do you work with them?"

The prosecutor, his face now reddening in anger, says, "The Is-lamic Republic of Iran has police in Kurdistan who spoke to the Kurd-

ish police, and they told them the three Americans were spies for the
United States! After the three got their mission in Baghdad, they met
with another intelligence official in Kurdistan. Then the Kurdish police *helped* them with their mission! *And* the Kurdish police found in
the area where they were arrested a map, books, and special documents that they were carrying!"

It is true that we hid a bag of stuff we were tired of carrying shortly
before we were arrested. In it was *The Shia Revival* by Ali Nasr, a
blanket, a tin of olives, and a bottle of whisky we planned to drink
around the fire with Shon that evening. I'm glad they don't know
about the whisky. The only thing that could qualify as "special documents" was Josh's journal with a goat face on the cover.

Josh and I lean in toward the translator as the prosecutor continues. "Our investigation revealed that the Kurdish police told the
three that they were nearing the Iranian border, but they ignored the
police. Everything I have stated was *admitted* by the three in their interrogations. They admitted to entering illegally. They admitted that
their country sends them around the world and pays their expenses
to spy for the U.S. and their allies. The fact that they lived in a neighborhood with Palestinian refugees is just one example of this."

Now I am angry — *what the fuck is he talking about?*— but my anger has a calm face. I have spent a year and a half preparing for this.
In the last couple weeks, Josh and I have been grilling each other with
every possible question we could imagine we might be asked.

"Furthermore!" the prosecutor shouts. "The testimonies of the
three during their interrogations contained inconsistencies, which is
proof that they were making up stories."

He ends his diatribe, nods to the judge, and takes his seat. Everyone I can see — the prosecutor, our lawyer, the judge, Josh — all of
them are expressionless.

Our lawyer gets up and presents a packet of documents to the
judge, which he says includes Sarah's defense.

"What is this?" the judge says, obviously annoyed.

"This is evidence of my clients' innocence I would like to present
to the court."

"You submitted your documents earlier and now you want us

to read all of this? You think I have time to read all this?" He flips through the papers momentarily, then sets them aside.

Then he asks us to introduce ourselves. He hands a piece of paper to the translator, who writes the question in English and hands it to me. The paper says exactly what the judge said: "Please introduce yourself." As in interrogation, I write in detail. I write what happened at the border, I write about my work, I write about my education, and I write about my life in the Middle East. Josh and I assume this is the only defense that matters. He needs to show he is critical of Israel, since we expect he will be branded an Israeli spy. And both of us need to show we are free thinking Americans who criticize our own government. Our best chance in court is to make them look like fools for imprisoning us.

When I am finished writing, the judge tells me to stand up and verbally state what I just wrote. I start reading from the paper, but he stops me and tells me to put it down. "Just summarize it," he says.

I recite as much as I can remember. The judge looks bored. As far as I know, the only individuals who can understand what I am saying are Josh, the translator, and our lawyer. "I've never had any contact with U.S. intelligence services in Iraq or in America," I say. "I am a journalist, and my purpose in my work is to expose and investigate the wrongdoings of my government, especially in the Middle East." As Josh and I practiced, I make sure to highlight my investigative articles on a U.S.-backed death squad in Iraq and military corruption.

The judge then reads me my charges again and asks whether I plead guilty or not. I ask him if I can speak to my lawyer first.

He shakes his head. "Just like anywhere else in the world, the court doesn't allow for that," he says. He is smirking. "But rest assured that your rights are protected. You can take as much time as you need to answer the questions."

When the judge asks Josh to introduce himself, he also stands up to give a verbal answer after he writes. He says, "I came to the Middle East about one week before I was arrested in Iran. After I finished getting my bachelor's degree at UC Berkeley in 2004, I worked as an educator. I was an educator in Oakland in 2005. I moved to Oregon.

I was an educator on the environment . . ." The judge tells him that is plenty and motions for him to sit.

We both realize what is happening at the same moment. Our verbal answers are only for TV. They just need enough for Press TV to grab.

The questions continue for both of us: Why did you spy on Iran? Did you meet with the CIA in Baghdad? Why did you come to our border after being warned not to? Why did you carry such a sophisticated camera?

As I write, the courtroom is utterly silent. For an hour, no one coughs. Every stir in a chair is heard. The silence adds to the purely bureaucratic feel to everything. At some point, the judge calls someone to the bench, hands him his keys, and asks him to move his car for him. He must have expected this to be over by now and didn't pay enough for parking.

At 2 p.m., the judge tells us we are done for the day. Our trial can't go any further without Sarah here, he says. He can't say when we will be back, but he assures us we will meet with our lawyer in the prison before the next hearing. Something about his self-assurance and his concern leads me to believe him.

As everyone gets up to leave, Shafii approaches us, but our guards intervene and try to rush us away. Shafii sticks his open hand into the air dramatically and turns to the judge. "May I shake my clients' hands? May I please be allowed to *shake* the hands of *my* clients?" The judge nods. With everyone in the room watching us, we shake hands. "Please tell Sarah I love her," I say three times. Then they pull us away.

83. SARAH

I've been awake all night, sitting on the couch in my friend's apartment, waiting for news of the trial. I sit with my laptop on my knees, sifting e-mails, texts, and Facebook messages. People from all around the world are sending prayers, songs, and pictures of candles they've lit to help them keep vigil on this night. As the sun rises, I can't believe we've heard nothing from Iran. No news reports, no phone calls from Salem, our lawyer, or the Swiss.

Soon, the house I'm staying at in Oakland starts filling up with dozens of people and dishes of hot food. My friends, family, and I huddle together on and around the couch and try to enjoy one another's company. Everyone wants to believe this trial will be the last hurdle. We read a poem by William Wordsworth, "Character of the Happy Warrior," passing it around from person to person. Shane and Josh had this one memorized; they used to recite it together, taking turns passing it back and forth like we're doing now.

> . . . *More skillful in self-knowledge, even more pure—*
> *As tempted more; more able to endure,*
> *As more exposed to suffering and distress;*
> *Thence, also, more alive to tenderness . . .*

Finally, the phone rings and Cindy tells me our lawyer's going to be on the line in twenty minutes. She just got off the phone with Ambassador Leu Agosti, who waited outside the courtroom along with thirty reporters for six hours, without being allowed inside.

For a year and a half, our lawyer, Masoud Shafii, has been systematically barred from defending us. He's never even been allowed to meet privately with his clients. But one thing the Iranian government can't stop our lawyer from doing is engaging the media. "The charge of espionage is irrelevant," he says in several interviews leading up to this session. "Sarah, Shane, and Josh do not fit the profile of spies." As for illegal entry, he continues, "there is no evidence of guilt in this respect either. Yet, if the court should decide they crossed the border, accidentally, then the time they have already served should be more than enough for this minor offense." Mr. Shafii is being more than generous. In fact, Afghanis and Pakistanis are arrested for trying to enter Iran illegally all the time. They're simply held overnight, perhaps charged a small fine, and sent back home.

The surprising thing is that Mr. Shafii insists on adhering so closely to the truth. In political cases like ours, it's common for Iranian lawyers to say their clients are guilty when they are not. They hope that if they play along, the government will eventually grow tired and release their clients; they also want to avoid ending up in prison themselves. Sometimes, this works, but, as in Zahra's case, there's no guarantee.

I tell everyone that I'm going to have to cut our vigil short, and I settle down for what I suspect will be a very long call. "Hello, dear families," our translator, Pari, begins. "Mr. Shafii is about to get on the line. He told me to tell you he is sorry to have kept you waiting, but he was exhausted after the court session and he feared he wouldn't speak coherently. He went home to take a nap and is now ready to answer all your questions."

Pari is one of the many miracles of our campaign. She and I became best friends instantly when we met in first grade in Los Angeles. I was a transplant from Chicago and she was the daughter of asylum seekers from Tehran. I taught her bad American habits like putting ketchup on eggs and even pulling down our pants when no one was looking and sticking our butts out to "moon" the sun. She taught me that people from faraway countries could be different and wonderful. Pari and I spent years of our childhood at each other's houses, then later lost contact in our teens. When she saw my face on TV in July 2009, she immediately tracked my mom down and offered to help.

Our lawyer's cheerful Farsi comes on the line. "How are you all? Did you sleep?"

Calls with our lawyer always begin with an elaborate exchange called *ta'arof* in Farsi, a form of communication marked by exaggerated kindness, generosity, and humility. For the first fifteen or twenty minutes, our lawyer deftly uses the art of *ta'arof* with each family member, inquiring about his health, making jokes, and disarming each of us with flattery.

"Alex," Shafii says, addressing Josh's brother, "I have been thinking of you and all the e-mails you have been writing me. I think you are an extremely good son, the very best, and when your brother, Josh, is freed, perhaps you will continue to make your mother proud by becoming the president of the United States."

"Oh, Shafii, you are too kind," Alex demurs. "I really don't think I'll want anything to do with politics once my brother is freed, but I hope you will come through with your promise to help me find a good, Iranian wife."

"Of course, Alex, of course. With all that hair on your handsome head, it will be easy. Speaking of handsome, where is Al? Are you there?"

"Yes, Shafii, I'm here," Shane's dad answers.

"How are you, Handsome Al?"

"Still handsome." Al laughs. "How are you, Mr. Shafii?"

Even in prison, certain guards, like Ehsan and Maryam, were the perfect embodiment of *ta'arof.* Once, I complimented Maryam's gold bracelets and she took them off to make as if to give them to me. When I took them out of her outstretched hand, she was shocked — in her cultural code it was my duty to refuse. Sheepishly, she told me she had to have the bracelets back or she would get in trouble — but when I was free, she would be happy to give them to me.

There's also a political dimension to *ta'arof.* Centuries of repression have led Iranians to use the verbal flattery as a way of passing information discreetly and indirectly. For instance, I began to notice Shafii's habit of repeating the same thing three or four times throughout the call ("The judge should know I work within the law"). At first, I thought he was just wordy and liked to hear himself talk. After talking to Pari, I realize the repetition, folded into flowery language and silly jokes, was intended for extra emphasis. The one thing I can never discern, though, is when this coded language is for our benefit and when it's intended for the Iranian government agents we assume are listening on the line.

By the time he's given individual attention to each of the eight to ten family members, we've all stopped squirming in our seats and surrendered to the ritual. As time-consuming as *ta'arof* may feel through an American lens, it has grown on us. Each one of us comes out feeling respected and seen. For eighteen months, our lawyer's been the main personal connection the families have had to Iran, so these calls are about much more than business — he's our link to Josh and Shane.

"Okay, now, that is enough *ta'arofing,*" Shafii cuts in, laughing. "Let's get down to our legal business. I know you must have many questions about the trial."

Masoud is sorry to break the news that he wasn't allowed to meet with Shane and Josh in private. He tells us they both were allowed to speak in their defense, but unfortunately no ruling was made, nor was a date for the next session announced.

When I finally get off the phone, I feel shooting pains in my legs and lower back. I realize I've been sitting on this couch for almost twenty-four continuous hours, yet I know little more than I did yesterday. I open my laptop and begin searching the web for news. After about fifteen minutes of reading empty news reports ("The two American hikers were tried today in Tehran"), a video suddenly pops up on BBC. Slowly, I draw in my breath and press Play.

The clip begins with footage of Shane and Josh walking into court, passing rows and rows of empty seats as they make their way down a narrow aisle. Josh's face looks strained but focused, ready to face whatever comes. Shane looks thinner than when I last saw him, but very much like himself, determined and resolute.

As I watch them walk through the empty courtroom, my eyes are drawn to the foot and a half of space between their bodies. It's a space just large enough for me to squeeze into. I've been waiting a year and a half for this moment. I prepared with Shane and Josh during the precious time we had together in *hava khori*. Now, sitting on this couch eight thousand miles away, I'm happy that no one else is around so I can allow myself one brief second of honest emotion. *I should be there,* I think. *That space between them is meant for me.* Despite all my work on the outside, I feel ashamed to be sitting here in comfort, out of harm's way. I'm ashamed I wasn't able to prevent this.

The video skips to a brief clip in which each of them addresses the court, giving their defense as to who they are and what they were doing on July 31, 2009. When the camera briefly passed over Judge Salavati, commonly known in Iran as "the hanging judge," I realize that Josh and Shane have no idea what a monster he is. This is the same monster that condemned Zahra to death, among many other political dissidents.

As painful as it is to tolerate a completely opaque, illegal court session parading as justice, I recognize that this hearing is still a victory. This clip makes Josh and Shane look like anything but spies. One is a grassroots environmentalist; the other a journalist who has investigated little-known activities of the U.S. military in the Middle East. Is anybody watching really going to believe the allegations against them? A few days later, the judiciary announces that there will be no judgment at this time. The next trial session will be announced "soon."

84. SHANE

Leaning back against a pillow, I hold three large pieces of cardboard. They are the most precious pieces of writing we have — timelines that span from 10,000 B.C. to the present. Whenever we are reading a book and come across the date of an event, whether it be the rise of the Tang dynasty or the invention of the bra, we crawl under the bed, remove a crumpled-up piece of plastic bag that is stuffed into a tube in the bed frame, and pull the pen out. (Our ink supply is precious — whenever we get low, we have to devise a new scheme to steal another pen from the guards.) Then we remove the proper timeline from under the carpet — there are three, scrawled across stretches of cardboard — and log the new date. We assume that someday, our cell will be cleaned out and these will disappear. This makes it all the more urgent to commit everything to memory now.

"When did the Montgomery bus boycott start?" I ask Josh, starting our twice-weekly history quiz.

"1955."

"When did Genghis Khan unify the Mongols?"

"1206."

"When did they convert to Islam?"

"1295, in Persia."

"When was the first known European contact with North America?"

"The Norse stayed in Newfoundland for a year in 1001."

"When was Al Jazeera founded?"

"1996."

"Name the CIA-backed coups that you know of."

"Mohammad Mossadegh was overthrown in Iran in 1953. Arbenz in Guatemala in 1954. Those two were democratically elected. Trujillo in the Dominican Republic in 1961. Sukarno in Indonesia in 1965. Lumumba in the Congo in 1960 — he was democratically elected. Salvador Allende of Chile in 1973 — also democratically elected. Umm, Noriega in Panama in 1989. That's all I can think of."

"You forgot the Dominican Republic in 1963," I say, scanning the cardboard. "It was another elected president. A social democrat, I

think. We don't have his name here. There was also another coup in
Guatemala that same year. There was something in Brazil in 1964—I
think that's all we have. Okay, when did the printing press come to
Europe?"

"Ahhh, Gutenberg's printing press was 1439."

"When did the poet Hafez die?"

"Aaaah," he says, squeezing the bridge of his nose. "I think it was
the 1300s?"

"In 1389. When and where was the first city founded?"

"Sumer in the fourth millennium."

"When did the Sassanids come to power in Iran?"

"224 A.D."

"Damn, you're good! Tell me what you know that happened in
1971."

"Hafez al-Assad became president of Syria. Bangladesh got independence from Pakistan. George Jackson was killed in San Quentin
prison . . . All right, that's enough. My brain is fried."

It's after dinner and this is the most socializing we have done all
day. When we do something like this, the tension seems to disappear. Earlier today, it felt like one of us could start screaming into
the silence at any moment. Now, as we interact, things start to feel
normal again. Sure, there is an underlying competition about who
can remember the greater number of dates—and Josh almost always
wins—but that tension is manageable.

We don't really have anything to say after we finish quizzing each
other. What can we say at this point that hasn't been said already? We
are both tired of discussing ideas from books, because we almost always take different sides by default and the act of debating lost its excitement at least a year ago.

Josh picks up the timeline that spans 1914 to the present, and I
study 10,000 to 1000 B.C. Instead of talking, I will memorize. Memory is a muscle that must be worked. When I get free, I want my
muscles to be stronger than they were when I came in. When I get
free, my mind will be an encyclopedia. That might make all this time
worthwhile.

85. SARAH

My lower back went out the day after the trial. For days I've been holed up in my friend's apartment, with half a dozen pillows propped up around me and papers spread out all over the floor. I'm typing a letter to Iran's Supreme Leader that Salem asked me to write, to be delivered through one of his envoys.

Our families, and a lot of influential people, have written dozens of such letters on our behalf—not even Senator Kerry's got a response. The only one that did was written by Jafar Saidi, a former Black Panther linked to me through a family friend. Jafar escaped a prison sentence in the United States by fleeing to Iran, where he lived and taught English for years. He was later arrested in Tunisia, extradited back to the United States, and is now serving a thirty-five-year prison term in Pennsylvania.

"The entire affair is the result of a grave misunderstanding," Jafar wrote in his letter to the Supreme Leader. "All three have extensive histories struggling with and supporting the *mustazafeen* [the oppressed]. They are not agents of the American, or any other foreign government."

"We want to thank you for your service to the Islamic Republic," the Supreme Leader wrote back a few months later. "We will take your request regarding the young Americans into consideration." Any response at all from the Supreme Leader is extremely rare, but what is truly amazing is that the highest authority in Iran took time to address the concerns of a former Black Panther, sitting in his prison cell.

The phone rings. I answer it, still typing my own letter.

"Hello?"

"Sarah, this is Sean Penn."

For a second I can't place the name; then it clicks. Sean Penn. *Sean Penn!* The super-famous Sean Penn, the actor from *Fast Times at Ridgemont High*, *Milk*, and *Dead Man Walking*, the one who's buddies with President Chávez of Venezuela — whose help our friends have been trying to enlist since this started.

"Sean Penn," I say, shaking my head and smiling to myself. "Hey!"

"Sarah, I heard from Reese Erlich that you wanted to talk to me. What can I do to help?" Reese is a journalist who went to Iran with Sean in 2005. Originally, Reese got in touch with me because he's a friend of the dentist who cleaned all the tea stains off my teeth when I first got out of prison. Reese interviewed me for a radio piece a few months ago and ever since then I've been hounding him to get me in touch with Sean. "Um . . . well, wow. It's great to hear from you, Sean, really great. Are you in L.A.? I'm hoping to fly there in the next couple of days. Perhaps we can meet up?"

"Yes, we can do that. I'll do whatever I can to help. By the way, have you been able to reach Chávez on this issue?"

Every time Shane, Josh, and I saw President Chávez on the news in Iran, we wondered why he wasn't involved. Iran and Venezuela share a strong alliance. In 2007 they established a $2 billion fund aimed at financing projects in the developing world "to help thwart U.S. domination." Venezuela was one of three member countries in the IAEA to vote against referring the Iranian nuclear file to the United Nations Security Council. Chávez supports Iran's nuclear program and even offered to supply Iran with F-16 fighter jets and suspend crude oil exports to the United States if it attacks Iran.

Perhaps equally as important, Chávez presents himself as a leftist progressive. How perfect for both of them to release the three *progressive* American hikers to one of the biggest thorns in America's side. In a realpolitik sense, both presidents would benefit from the propaganda this would generate. How could Ahmadinejad afford to say no?

A few days later, I'm in a cab on my way to meet Sean for dinner at Madeo Restaurant in West Hollywood. Sean's date is the actress Scarlett Johansson, who looks exactly as gorgeous in person as she does on camera. I slide into the booth next to them and we begin to talk.

I like Sean immediately. Something in his demeanor, his quick eyes that give you their complete attention only to suddenly take it away, gives me the impression of a man who carries more than his share of the world's burdens. After talking for just a few minutes, I can tell that Sean gets my intensity. He gets that nothing will stop me, and I get that he respects that.

"Sorry, Sarah, I don't mean to interrupt you — I just can't believe I'm looking over your shoulder at Elton John," Sean says, breaking the spell.

I turn around to look and Sean's right; Elton John is sitting a few tables away from us. There are other famous people in the restaurant that I vaguely recognize and know I should be able to identify but I can't. I've always been a little out of the loop when it comes to mainstream culture. I wasn't allowed to watch TV much as a child, and even though I love film, I've never felt any kind of fascination with the personal lives of the actors behind the roles they play. I don't feel nervous or starstruck. Actually, the great thing about meeting famous people is that you feel like you already know them. It's funny because that's something I've grown used to hearing all the time from strangers who've been following our case.

I ask Sean about his work in Haiti, about the 55,000-person refugee camp he currently runs. "What made you decide to do it?" I ask. "Why did you become an activist?"

"I'm not an activist," he says. "I just do stuff because it needs to get done."

"Perfect," I say. "That's exactly what I need, to get this done."

After dinner, Scarlett and I squeeze into the front seat of Sean's beat-up pickup. He speeds the truck out of the basement parking lot, jumps the curb, and races away so fast that the paparazzi can't catch up with us. One good picture of him and Scarlett, he tells me, could be worth fifty thousand dollars. "And those bastards don't deserve it," he adds.

Sean asks me about what happened right after we were captured. I usually don't talk about those days, mostly because it feels like a long time ago and the present moment is much more urgent, but I decide to go into it, describing being handed off between different groups of men in suits, driven out of the city at night, seeing the man in the front seat cocking his gun and thinking we were going to be murdered.

Scarlett asks me if I want a cigarette and I accept, the first I've had in years. Speeding through the streets of L.A. with my new friends, I think about how incredible it's going to be to celebrate with Josh and Shane when they get out. It's going to be hard to share them though. I can understand why Sean and Scarlett try so hard to keep their romance and personal lives to themselves. I feel the same way when people want to see the engagement ring Shane wove for me in prison or ask me to recount the story of the day he proposed to me. I can't wait to be anonymous.

The next morning, Sean and I get to work. First, we ask Noam Chomsky to coauthor a letter with Sean, imploring President Chávez to intervene on our behalf. The letter will be hand-delivered to Chávez along with a packet of articles and essays written by Shane, Josh, and me, proving our progressive credentials. Amazingly, we're even able to dig up an unpublished essay in support of the Bolivarian Revolution written by Josh.

Sean checks in with his friends at the State Department and they say Secretary Clinton has no problem with our initiative. We soon get news from the Venezuelan foreign minister that Chávez wants to help. I ask Sean if he'd be willing to fly down to Venezuela and meet with Chávez himself. Incredibly, he agrees without hesitation. A week later, I get the text I've been waiting for:

> Sean: I'm in Venezuela. Met with foreign minister. I do not want
>   to create any false hopes, but it seems they are becoming quite
>   focused on this, and conversations with the other side are good.
>   I've gotten a lot of winks and smiles intended to assure me it
>   will work out well. Proof will be in the pudding.
> Sarah: Really, this is incredible!!!!!!
> Sean: I only have one mantra, two humans in a frighteningly
>   dangerous cage. One, also longing for his love . . . could be me,
>   or my kids. My internal impatience is fueled.
> Sarah: Thank you for putting yourself in our shoes.
> Sean: I took the liberty to tell him that many in their families are
>   being torn apart. Chávez is a very sentimental man. He took my
>   meaning that this must happen quickly.

A few days later, I hear back from Sean that a phone call has taken place between Chávez and Ahmadinejad. Chávez asked for Shane

and Josh to be immediately released as a personal favor. He even offered to fly to Iran and pick them up himself if necessary.

"Normally I don't intervene in our judiciary, but for you — my
brother — I will do it," President Ahmadinejad replied.

86. JOSH

It is the beginning of April and Day 7 of our hunger strike. Shane and
I just played rock-paper-scissors, and he won. That means that he'll
go to *hava khori*, and I'll stay in the cell and pretend that I'm too weak
to go. I wish he'd volunteered to stay in, given all the times that I went
to *hava khori kucheek* for him and Sarah. But I wasn't doing that for
trade, so I have no right to ask for compensation now. Our friendship
is increasingly becoming about "rights" instead of giving. Anyway, I
agreed to play rock-paper-scissors, and Shane hates when I even ask
about changing an agreement.

I've plenty of energy for *hava khori* and would love to go, even
though I'd have to act weak and sit down pathetically in front of the
security camera. There is something special about the open sky, especially during springtime when I smell the wildflowers blooming
just beyond the walls.

I have energy because I'm not actually hunger-striking. We
stocked up on food in the days prior to this hunger strike. We requested extra flatbread, then sealed some in plastic bags and dried
more into chips. We hoarded butter, cheese, milk, yogurt, nuts, juice,
crackers, oranges, apples, bananas, carrots, cucumbers, cabbage, tomatoes, onions, tahini, jelly, and honey. The night before we started,
Ehsan happened to be working. He's been working here again and
he's started talking to us whenever he's on duty. He brought a plate
of five extra hamburgers — we stuck them in the fridge. The excess
white rice from last week stores nicely in the freezer section of our
fridge. We can even make tea with the hot water from the shower.
We keep the food hidden, though, because we want them to think we
aren't eating.

Before we started "fasting," we agreed to ration our bread, rice,
butter, and other items to prepare to be without meals. We allotted fifteen days' worth of food because Dumb Guy is on vacation for

Nowruz, the Persian New Year. As the days progress, the mold has colonized our bread. I've preferred the peace of mind of saving more for later — even if I lose some pieces to mold. Shane has wanted to eat more bread. So instead of sitting anxiously while Shane eats our collective bread, we decided to privatize it.

On Day 1, almost everything was collective. One week later, every morsel of food is privatized. It feels like a failure, like I am not being selfless enough. Privatizing avoids the difficult conversations required to agree on our consumption. It's a less generous way to think and to live. With privatization, I watch Shane vigilantly as he divides the yogurt that we share at lunchtime.

It's odd, because Shane and I have discussed cultures that live in scarcity, like the rebels he visited in Darfur and the people in places I've been in the forests of India, and we talked about how they share food partially *because* they have so little. (We've also calculated our calorie intake while "fasting" and we actually consume more calories than these people.) Yet we are doing the opposite. When we felt abundant, we shared; now we don't. But splitting the yogurt makes life easier.

With all the food we're consuming, we've struggled with how to deal with the trash: orange peels, cracker wrappers, yogurt containers, etc. First we took down the onion plant on the windowsill and stuck our organic waste under the topsoil. That filled up very quickly. Then we sealed our trash in a bag and stuffed it under the carpet, but soon a mound bulged from the floor. Next, we double-sealed bags of garbage and hid them in folded sweatshirts, but it still stank. Our last strategy has been to throw the organic waste down the toilet and the wrappers under the carpet.

That didn't work at all. The janitor had to plunge our toilet because I didn't shred the orange peels into small enough pieces before sending them down the drain. On the third day, our bathroom clogged and flooded with orange peels and our feces. I got down on my hands and knees and used a plastic spoon to remove each orange peel, one at a time, from the flooding toilet water before the janitor came so that he wouldn't see our waste-disposal scheme. Shane wasn't going to scoop up the orange peels. He had told me to shred them smaller when I disposed of them.

When guards come by, I pretend to be bedridden — instantly summoning my memory of how I felt during real hunger strikes. It feels like we're losing this hunger strike, but that seems appropriate. We started it almost mechanically, without any zeal.

Our threat to strike every thirty days is more about our power struggle with the interrogators than out of any present need for connection to the outside world. Letters don't nourish me like they used to. They are still sweet and wonderful, but I haven't cried reading them in months. I don't read them as voraciously as I used to. Sometimes I don't even read Shane's family's letters until the next day. This hunger strike is only for Shane's sisters' letters. I wanted to demand Alex's also. Though I receive some of his letters, some of them are missing. Shane insisted that we keep the strike focused only on the most salient censorship, and I conceded.

Deception pervades our whole effort. Nurses come to check on us, but we refuse medical care because a blood pressure check would show that we are faking it. Perhaps if our hearts were more in it, if we had more intention around this strike, I wouldn't have clogged the toilet with orange peels. Maybe Shane wouldn't have been caught carrying a thermos from the shower filled with hot water. Simply by eating, we're taking ourselves out of the right mindset.

Shane and I think the guards are skeptical. Shane asks me if I think guards have some instinctual, animalistic knowledge of when other humans are starving. He knows I'm apt to believe in the quasi-magical capabilities of the unconscious. I do think the guards know, and I fear we're losing credibility daily.

By Day 18, I feel like moldy bread crumbs at the bottom of a clear plastic bag. As much as I lack food and nutrition, I crave *hava khori*. We actually ran out of food and haven't eaten in three days. Every few days Shane asks me if it's okay if we extend this hunger strike for his sisters' letters. He would do the same for me, I tell myself whenever the meal cart comes around, taunting us with warm, aromatic sustenance. Plus, it's important that we don't let the interrogators win.

Dumb Guy shows up and I follow him to the interrogation room. I put my hands on the desk and slowly lower myself into a chair across from him, exaggerating how weak I am. I go alone to the interrogation room and do the negotiating. Shane pretends to be bedridden and ill. Dumb Guy knows Shane has had recent medical problems and we want to scare him. Dumb Guy sits on the opposite side of the desk, watching attentively. He looks around. "Where's Shane?"

"He's bedridden," I say coolly. "When he stands up, he gets lightheaded and has bad headaches."

"Our letters, our letters . . ." I give a dramatic pause and close my eyes. This is our last chance to save face. Shane and I already decided we're going to eat tomorrow even if he doesn't bring our letters. I pretend to suddenly remember my train of thought, "We need our letters. Shane's sisters' letters. Both Nicole's and Shannon's. Why don't you bring them?"

Dumb Guy speaks gently. "Have you eaten?" I shake my head and keep my eyes down, half covered by my eyelids.

"Seventeen days," I say, telling myself I'm not actually lying because I haven't uttered a complete sentence. For some reason, it matters to me that I don't speak a lie while I do everything in my power to deceive him.

Dumb Guy's attention shifts to the doorway. My interrogator enters the room with a tray of food — rice mixed with lentils and a few dates. A couple kebabs and ketchup packets sit patiently on a separate plate next to two bowls of yogurt mixed with cucumber. He slides two manila envelopes toward me. "Will you eat?"

"I will if Shannon's and Nicole's letters are in here." I open Shane's envelopes and sift through the stack of letters, sorting them into piles by sender. When I get through the letters, I tell Dumb Guy that I need to check with Shane first. We've rehearsed this drama. Escorted by a guard, I walk back to the cell, letters in hand.

I cough and make loud footsteps to make sure Shane knows I am coming down the hallway. His neck is barely elevated above his pillow. His water bottle leans against the edge of his mattress by some dirty tissues and the remote control. The guard stays in the hallway and doesn't speak English. "Okay, Shane," I say. There is no triumph in my voice, just relief. "They brought letters. There are some from everyone. But I gotta go back and tell Dumb Guy we will eat. I'll be back shortly with food. It's kebab."

I chew my kebab in front of Dumb Guy. The taste of warm protein immediately soothes me. Dumb Guy takes advantage of his captive audience.

"Josh, you have to help Shane. He does not know what is best. You know this is very difficult for him. Can you see that?"

"Yes, it is difficult for Shane."

"Shane is not very religious. You are Jewish. Jews are more religious than Christians. You have to help him. He needs to pray — you can teach him how." I bite my tongue at the absurdity of his statements, and he continues. "He is too angry. That is why he is having his stomach problems." For a moment I think about all of Shane's recent trips to the doctor and how he has less stomach pain when he meditates. "I don't want Shane to know certain things. If he knows about everything, it will make life more difficult for him."

"You censor our letters!" I blurt out.

He always insists that he gives us *all* the letters that he receives. We've never known if he's lying or if it's his superiors who withhold our letters from us. I bite my tongue again, hoping he'll continue to betray himself.

"No. I told you. I don't censor your letters, but I worry about Shane. Do not tell him about this, but Sarah is not doing so well. She is having a hard time."

In her letters, she wrote about the difficult conference calls and personality clashes, but what does Dumb Guy know that I don't?

"Shane is angry. If he knows about this, it will only make life more difficult for him. Josh, please do not tell him anything I told you."

Dumb Guy thinks I would keep a secret from Shane. He seems sincere — that is the most disturbing part. I think he is also trying to use the fact that I went on hunger strike for Shane's sisters' letters to drive a wedge between us. Shane and I have our differences, but we'll always stand together against Dumb Guy.

As I slowly rise to leave, I assure Dumb Guy that if he is really worried about our health, the best thing he could do is to stop censoring our letters. They will make us healthier — not least because we won't need to hunger-strike for them again in thirty days.

87. SARAH

*4-15-11 Last night, I had to call the Internet service help line at the
hotel where I'm staying. I've always rushed through these types of
calls, but I found myself totally content to stay on the line with a
friendly voice. It's not often that I admit that I'm lonely. I know that
you guys want me to appreciate my freedom even though you don't
have it yet. I appreciate it in small ways more than anything, the
wind on my face, a new human being, a song. The initial incredible
beauty of it has diminished without the two of you. I was recently
diagnosed with PTSD — I've informed our judge that this is why I
can't come back for the trial. I don't want you guys to worry about
me. A lot of my symptoms, emotional detachment, insomnia, hy-
pervigilance and hyperarousal, are helping me cope. It just amazes
me that the families have been able to keep this up for so long.*

For the last two months, the campaign has been a black hole. We
pour everything we have into it, every ounce of hope and energy,
every waking minute, yet nothing ever seems to come out. I feel like
I've been through fire, like my skin has hardened into a thick, protective shell, but underneath I've never felt so vulnerable.

I talked to Salem last week and the news was not good. He said
there was a meeting with all the Iranian authorities involved in our
case. He said that there were "certain elements" present that would
like to do Josh and Shane harm, but that after much discussion, they
realized that "all they could do was continue to deprive them of their
liberty."

I hate all the euphemisms and double talk. They are already doing
them harm. What they mean is execution — hanging the innocent
"hikers" by their necks. The fact that Josh's and Shane's execution is
on the table, even if only for fanatics like Judge Salavati, makes it impossible to deny just how high the stakes are and just how far off freedom might be.

That's why we can't slow down. The most important task I have
is to continuously keep the urgency alive for key players like Iraq,
Oman, and Venezuela — though lately, each of these tracks have been
hopelessly riddled with setbacks.

The president of Iraq, Jalal Talabani, traveled to Iran last month, ostensibly for the Iranian New Year, Nowruz. Ahmadinejad persuaded Talabani to take time from his busy schedule to attend the celebrations, under the pretext that he'd leave Iran with Shane and Josh in tow. When Talabani got there, Ahmadinejad avoided him for days. Finally cornered, the Iranian president produced an executive order addressed to the Revolutionary Court, ordering the charges against Shane and Josh to be dropped. He said it would be delivered after the holiday. Talabani couldn't wait around another three days — so he left an envoy in Tehran to bring Josh and Shane back to Baghdad. After the holiday, his envoy was snubbed.

Then there's President Chávez. A month after agreeing to release Josh and Shane as a personal favor to his "brother" in Venezuela, Ahmadinejad asked for just three more months to get it done. When those three months expired, Chávez said he was going to pull a Houdini and show up in Tehran unannounced, asking for what he'd already been promised. Everything seemed to be falling into place. Then, we find out President Chávez is in the hospital battling cancer. Then he's seeking treatment in Cuba. "Things are on hold while Chávez heals from surgery," I hear from Sean. "He won't be able to travel any time soon. I'm so sorry, Sarah."

No one's saying they're giving up, but to me they seem all too content to chalk up these setbacks to diplomacy-as-usual. It's impossible to know which, if any, of these efforts will bring Josh and Shane home.

The families ask the State Department for a phone call with Hillary and she agrees. With nothing else promising on the horizon, we broach the never-popular subject of a clandestine prisoner exchange, in particular involving Shahrazad Mir Gholikhan.

"The State Department has already passed on our positive recommendation to the White House on this issue. I recommend you take it up with them strongly," Hillary says.

"We already have, Madame Secretary," Alex responds. "Many times. They tell us to take it up with you."

"Well, we can go back to the Department of Justice and push them, but really this decision lies with the president."

All this finger-pointing reminds me of when I was still at Evin.

Every time Shane, Josh, and I asked our interrogators to justify our treatment, they would always pass the buck — it was the judge's decision, our government's fault. They never assumed responsibility themselves.

"Madame Secretary," I begin, "there's something I've wanted to ask you for a long time."

"Yes, Sarah."

"Why did the FBI lose interest in our case after the first few weeks? And why hasn't the CIA been involved? They're usually the ones who arrange prisoner swaps, right?"

"Generally, yes, that's correct."

"Isn't it true that without CIA or FBI participation, this will never get done?"

"Sarah," she says, then pauses. "I'm sorry, but I can't speak for any other agency, much less their motives."

Usually, when Americans are held hostage or kidnapped in another country, it's the FBI that gets involved. They've saved journalists from being executed by the Taliban and even orchestrated a prisoner swap of U.S. and Russian spies in 2010 while I was still sitting in a cell in Iran. Recently, the CIA got Raymond Davis out after he killed two people in a gun battle on the streets of Pakistan. Obama called for Davis's release, citing "diplomatic immunity." The thing is, he wasn't a diplomat; he was an ex–CIA agent and a contractor for Xe, a private security firm formerly known as Blackwater. Less than a month after the murders — while thousands of Pakistanis protested in the streets, calling for his execution — the Pakistani government was persuaded to accept blood money for the victims; then Davis was acquitted and sent home. Why won't the CIA at least try to do for Shane and Josh the same that they do for murderers and real spies?

For the last six months, the representatives of the three main countries involved in negotiations with the Iranian government on our behalf — Iraq, Oman, and Venezuela — have all reported directly to us. We have more information than our government about our case. On our weekly calls with the State Department, more often than not *we* are the ones updating *them*.

Not to mention that the people from the State Department's Iran

Desk assigned to our case have never even been to Iran and don't even speak or read Farsi. The State Department has been successful in convincing many countries (like Brazil, Qatar, Turkey, and Canada) to raise our case with the Iranian government, thereby keeping diplomatic pressure strong. They've also tried to be creative, like the tweet State Department spokesperson Philip Crowley sent to President Ahmadinejad on his birthday, asking him to release Josh and Shane. "Your fifty-fourth year was full of lost opportunities," he wrote. "Hope in your fifty-fifth year you will open Iran to a different relationship with the world."

The State Department has also missed a lot of opportunities. Just weeks after my release, I met with Imad Moustapha, the Syrian ambassador. He told me that Bashar al-Assad, the Syrian president and one of Iran's closest allies, had already expressed his willingness to intervene on our behalf. They were just waiting for the U.S. government to ask them. In fact, our families asked Secretary Clinton to do just that and she agreed — but failed to follow through.

I have no doubt that the United States wants Shane and Josh freed, but I've come to believe that they are willing to risk almost nothing to make that happen.

On a State Department conference call leading up to a huge press conference we're doing with Muhammad Ali in DC, organized mostly by Josh's brother, Alex, we decide it's time to stop asking their permission and use what little leverage we have.

"You know," Shane's mom, Cindy, begins, "reporters are constantly asking us what our government's doing for us. If they ask us at this press conference, we're going to give them an honest answer."

"I understand your frustration, Cindy," the Iran Desk representative, Michael Spring, replies, "but you realize that if you publicly criticize the U.S. government, you'll be sending the wrong message to Iran."

He knows there's a lot to be critical of. Instead of addressing those concerns, Michael counters our threat with another, more sinister one. He suggests that criticizing the U.S. government's inaction could embolden Iran, that we'd be doing their bidding. As long as Iran is getting something out of this, this logic continues, they are going to hold on tight to their last two hostages. Michael's playing on our deepest fears that something we say or do could make the situation worse for Shane and Josh, but we're used to that. The Iranian government, through Salem, Livia, and other diplomats, has given us similar warnings about the consequences of criticizing Iran.

With respect to diplomacy, nothing of consequence has happened since my release. Josh and Shane have already been in Iranian custody much longer than any other Westerners or dual nationals in recent history. Two German journalists were released last October after four months in prison. The American hostages in 1979 were released after 444 days, a mark Shane and Josh surpassed six months ago. We've been waiting for court for over twenty months—if the United States was going to do something substantial, it would have already done it.

At times, the stress is almost unmanageable. Last week, I broke out in hives all over my body when I found out some campaign funds we've been expecting didn't come through. I feel defeated, as though I might sink down into this abyss of bad news and never get up. I thought I was going to be able to make a difference—that having me on the outside to fight for them would make a difference. Sometimes I wish I were back in prison, where everything hurt but I was responsible only for myself. But that was what I hated the most about prison—nothing I did mattered. I wish I could rest, have faith that this will all work out on its own. But I can't. I don't. It won't.

Last week, a stranger in a restaurant asked the waitress to pass me a clandestine note scribbled on a receipt. "Just remember we're all with you," he wrote. Even at my lowest points, there's always someone or something to remind me I'm not alone out here. With so much love and support I know I'll always find a way to keep going. No matter what happens.

88. JOSH

Together, we construct a spice rack by sticking plastic spoons into the radiator and we make a new bookshelf with a food tray. We dust the TV, polish the cell door, rearrange the photographs of friends and family, and tear world maps out of books and tape them to the wall.

We reorient the beds so it's easier to exercise. For the first time, we borrow an electric vacuum, and Ehsan allows us to trade in our dirty wool blankets for clean ones. We make the room *ours* — both mine and Shane's — more than it has ever been.

But, as we check off more and more items from our to-do list and the days pass, the specter of going back to the tedious routine of reading, eating, and exercising encroaches. Tensions that we brushed aside during the hunger strike resurface.

When he turns on the TV to check the time, I habitually crack my toes. When I turn the pages in my book loudly, he snorts. Through body language we engage in a continuous, and half-conscious conversation. When I try to talk about it with Shane, he says he doesn't even notice the dynamic.

I need to take a walk, drink a beer, call a friend. I meditate, write in my journal, and read, but all I need is the space I don't have.

As a way to create space, I suggest that we partition the room into two halves. Shane agrees easily.

We sit down next to each other to unstitch a hem on one of our new blankets. As we work, we laugh at ourselves and joke about how the guards will interpret our partition. Then we hang the fabric between two nails on the wall to make a clothesline across the cell. Next, we drape a blanket over the clothesline and place it so that we can't see each other from our regular sitting positions.

A guard peeks into the cell. He sees the partition and bangs his fists together, asking in Farsi, *"Are you guys fighting?"* We chuckle when we hear this. We're not *fighting,* but I love not always being within Shane's peripheral vision. I love not having him in mine. By dividing the room in half, I feel more free.

89. SHANE

We are in *hava khori kucheek* and I am cutting Josh's hair. We've been asking the guards to let us do this for two weeks, and now that we finally have an electric razor, it is taking forever. Josh's hair is so thick and this machine is so old that I can't just mow through it. Every time I push into a clump of hair, the motor slows and almost dies. We tried asking for scissors, like we always do, but the guards won't allow it.

Instead, they just peek in from time to time and giggle at us in our shirtless state, covered as we are in little specks of hair.

Josh has Mike Davis's *Planet of Slums* and he is testing me on random facts as I slowly chip away at his hair. What are the ten largest cities in the world? What are the largest slums?

In the adjacent bathroom, we hear the shower turn on. We both dart over to the window. Josh pushes it open slightly. "*Salaam*," he whispers. The window opens fully. "*Salaam*," says the man on the other side, smiling. He is naked and shrouded in a cloud of steam. We exchange pleasantries, and he tells us he is a Kurd, from Iraq. He speaks only a smattering of Arabic, which, between that and Farsi, allows us to communicate a little beyond the basics.

He knows who we are. "Aren't you the one who is married?" he asks me.

"I am not married — we are engaged," I say.

"Her name is Sarah, right? Where is she?" he asks. I tell him she was freed.

He nods. "That's very good," he says.

I ask him why he is here.

"I'm a member of al-Qaeda," he says matter-of-factly. "I am from a village near Ahmed Awa, where you were captured," he says. "My cellmate is a journalist, also Kurdish, and he talked to some people there. He said Iranian agents paid some Kurds there to tell the Iranian media you were spies." I'm not sure I believe this — though it could explain the Press TV story my mom told me about when she visited in May, when Sarah is reported to have told an Iraqi Kurd she was Iranian and on her way home. Still, I am impressed by how angry he is about it.

"Everyone knows you guys are innocent. You shouldn't be here. This regime is playing politics with you." We talk about his case. He hasn't been to court. He doesn't seem to expect that he will ever go to court.

"What is your religion?" he asks.

"Christian," I say, lying.

"What about him?" he asks, pointing to Josh.

I translate to Josh. I always let him take that question.

"Yahudi," he says. The prisoner nods.

"Jews and Christians are our brothers. They are People of the Book." He says this in a tone a Mustang owner would use to concede the merits of a less exciting but admittedly reliable vehicle. "But Islam," he says, drawing out the word and smiling. "Islam is light!" He is beaming. "Why don't you convert to Islam?" The limits of our shared language are starting to show. "With Islam, you will sleep better at night," he says, putting his head on his upturned hand as if it were a pillow.

"Inshallah," I say, my usual response to conversion attempts. It's such an easy, noncommittal statement and it's hard for a religious person to argue with.

He gives us one more beseeching, almost motherly look, a last invitation to cross to the side of Islam. I smile by way of declining politely. "God willing, you will be free soon," he says.

"You too," I say. He waves goodbye and closes the window.

90. JOSH

I wake bleary-eyed to the sound of the rattling breakfast cart. I look over toward Shane. He looks like he's been up for hours already. Our next hearing is set for today, May 11. When he visited last week, my interrogator wouldn't admit that we were scheduled to have another hearing until I let on that I already knew about it from the letters.

It's been a tough week. I'm hopeful and nervous. Everything feels a little tense. I have a little too much energy at *hava khori.* And I barely sleep at night. I wasn't so nervous before our first trial, but for some reason, this one has me on edge.

The news occasionally reports executions at Evin Prison. I'm not worried about that prospect. We have too much international attention. I'm expecting a twenty-year sentence, maybe twenty-five. They'll prance us around the media and boast to their people about how tough they treat conspiring foreigners. Then they'll hopefully release us afterward and boast about how compassionate they are.

In letters, everyone seems to be almost expecting our release with this trial. Reading and rereading Sarah's March 14 letter, I gain even more hope. In it she details how foreigners have been released from Iran:

> Two German journalists . . . Twenty months was then commuted to a $50,000 fine . . . A French woman . . . released on bail . . . convicted to two five-year terms, which was immediately commuted to a $285,000 fine . . . Three Belgian bikers were detained three months in 2009 . . . A dual-national woman sentenced to eight years . . . detained for one hundred days.

Compared to my mother, I feel like a cynic. But Shane thinks I'm too hopeful. I watch him repress his hope. It's been a while since I allowed myself to dream of freedom, and the dream feels good. I try my best to follow the advice of *The Myth of Freedom*, a Buddhist book my friends sent me: allow my hope to exist, acknowledge it, and observe it, but do not get attached. It says that the *attachment* to hope — not the optimism itself — creates suffering. But it's difficult to not be attached to freedom.

I eat my breakfast and I wait. At one point, I take yesterday's chicken wishbone off the radiator and present it to Shane. We tug at different ends. Its head flies off, neither of us winning.

No one arrives to take us to court by 11 a.m. "We still have a few hours. They may still come," I tell Shane, though I know he doesn't believe me.

At 2 p.m., I say, "It seems improbable, but it's not yet impossible."

The guards won't tell us anything.

At 5 p.m. we go to *hava khori*. "Maybe there was a political breakthrough," I hypothesize. But of course, I have no idea why we weren't called.

91. SARAH

"Okay, this is too much," Josh's mom, Laura, says. *"Too much!"*

It's May 11 and we're on a family conference call. We just found out from our lawyer that the trial was canceled by Judge Salavati because Shane and Josh "didn't appear in court." This is the exact same excuse given by the court for canceling the trial last November. The difference is that not a single person outside of Evin Prison has seen or heard from Shane or Josh since February 6, more than three months.

"How can they not even show us the boys?" Laura continues. "They could be hurt; they could be hospitalized. *How can they not even let them talk to their mothers?*"

This is our biggest wake-up call yet. We have no way of knowing for sure if Josh and Shane are even alive. At the very least, we fear they are hunger-striking. It feels like we're back where we started, that trial dates could be canceled and postponed for years to come. This time, the Iranian government went too far, and our families are ready to fight back.

"I think it's time we change our strategy," Cindy says. "We need to get their attention."

"I feel certain they started hunger-striking when they found out the trial was canceled," I say, then add, "If I were still there . . . that's what I'd be doing."

"We've been talking about the idea of starting a public hunger strike ourselves," Cindy says, then adds, "I think the time is now."

"I agree, Mom," Shane's sister Nicole says. "I think the moms should start it — that will be big in the media." Shane's other sister, Shannon, agrees.

"We can also have Sarah start talking about the prison abuse, sexual harassment" Alex says, then adds, "That kind of stuff really embarrasses the Iranian government."

"Let's look at one item at a time here," Paul Holmes, our media advisor, interjects. "First, I say we vote on the hunger-strike idea. We can start with Laura and Cindy — they can fast for a few days and do interviews. Then Sarah can join, do more media, then anyone else in the families can continue to join as others drop out. We can call it a *rolling hunger strike.* I'm confident that will get coverage."

Paul takes a vote and we unanimously agree on the rolling hunger strike, deciding to pass the torch indefinitely until we're convinced that Shane and Josh are safe. Compared to our usual drawn-out battles, this decision is one of the easiest we've made yet.

92. SHANE

We have five minutes. This will be our third phone call since we arrived here. It's the end of May, six months since we have made a phone call, six months since I wrote down Sarah's phone number and committed it to memory. This is my chance. As I dial her number, I know it will probably never happen again. You can only break the same rule once. They told me I could call family and, to them, family doesn't include Sarah.

The phone rings and keeps ringing. How could it be possible that she doesn't answer? The answering machine picks up. I start to speak. "Sarah, it's me. I can't believe you're not there! I love you so much and I just want you to know that I'm good. I'm healthy. I'm strong mentally, emotionally . . . Things are good and things have really gotten a lot easier now that you're free." My voice is cracking. *Keep it together. Don't break down on the phone right now,* I tell myself.

As I talk, Josh's interrogator, standing next to me, is starting to look around nervously, as if for help. Dumb Guy is down the hall, putting Josh in our cell. Josh's interrogator yells to him. I keep talking hurriedly, blocking them out. "I sound emotional now because it's amazing to hear your voice. I love you so much, Sarah. You are incredible." Dumb Guy is striding toward me. "I can't wait to see you. I can't wait to hear your music . . . Please send music. I love you, baby. Bye."

"What is this?" Dumb Guy says. "What are you doing?! We told you to call *family,* Shane. *Family!*"

"I know, and that is what I did. I called family."

"You broke the rules, Shane. You may never call Sarah again. Do you understand?"

"I understand."

A part of me feels victorious and a part of me feels gutted.

93. SARAH

5-24-11 I usually leave my phone on vibrate because of the remote chance that you might call while I'm in acupuncture. This session put me into a deep trance and the phone must have dropped out of

my hand. When I walked out of the session and got your message,
it felt like I was still dreaming. Your voice shot through me like cold,
dark water. I stood still for a long time, watching light rain fall on
a grassy hill, in utter amazement at your strength and resilience.
I heard so much life in your voice, I felt instantly so much closer
to you and our future together. I'm a lot more stable out here. I
don't have highs and lows really, but I also don't feel much joy. Your
phone message gave me the permission to feel joy, real joy! The fact
that you're able to give me that, after all you've been through, as-
tounds me. I surrendered right away to the fact that I missed your
call, knowing how much it would mean for your dad, sisters, and
friends to hear your message. Now I can listen to it every day until
I see you again. Thank you, baby.

It doesn't make sense that I missed the call; for the last ten months
my phone has never left my sight. I sleep with it plastered to my head.
I often wake up several times a night to double-check that no one's
called and check my e-mail. After Shane left me a message, he called
his mom. In the midst of their short conversation, Shane told her
they had "time for their tears." This time they were hunger-striking
for seventeen days. I never dreamed they would take it this far.

In a sense, the phone calls are our first tangible win, at least since
I was released. When Laura and Cindy went public with their own
hunger strike two days ago, Shane and Josh were allowed a call home
the next day. The Iranian government gave in, so we now have proof
that media pressure yields results. Now, it's up to us to show them
that two five-minute phone calls will not be enough to silence us
again.

Our worst fear has always been that something we do on the outside will put Shane and Josh in greater danger. In the past, that fear
has immobilized us. Yet, the opposite just happened — our bold move
on the outside resulted in something positive for them. As small as it
may be, it was a victory.

I realize that my mom got to the same point almost exactly a year
ago. She got fed up with waiting around for other people to get me
out of prison. She saw my breast lump and solitary confinement as
an opportunity and, with the families backing her up, jumped on it.

That strategy worked. When I asked the judge why I was being released before Shane and Josh, he said, "You're a woman — in Iran we treat women well. Also, you've been in solitary confinement."

The Iranian government is very sensitive about its atrocious human rights record. With a high-profile case like ours, they put a lot of work into making it appear that Shane, Josh, and I have been held under good conditions. We realize that in addition to the incident of Shane's beating, the only thing we can use for Josh is the inappropriate behavior he experienced from one of the guards, who habitually made groaning noises at him, even when he was in bed at night. After it happened, the three of us discussed it in *hava khori* and Josh was convinced he was being sexually harassed. We all agreed he should complain to Dumb Guy.

The families and I decide I will fly to London and break the news of prison abuse and harassment on BBC's *Hardtalk*, both the English and the Farsi version, since this is a show with an international reach that's also popular with Iranians.

Watching the families struggle with these tough decisions reminds me of the long, drawn-out debates Josh, Shane, and I often had in prison. I think about when Josh and Shane stood up to AK, when Shane told Dumb Guy never to touch him again, or when I slapped Maryam back and told her I wasn't a child. By speaking truth to their lies and manipulation, we were able to take back some of our own power. We eventually realized that if we didn't stand up to our captors, they would walk all over us. After nearly two years of dedicated but cautious advocacy, our campaign is coming to the same conclusion.

Many Iranians and Iranian Americans think our families have been too nice since the beginning and that's why our ordeal has dragged out so long. They say the Iranian government will continue to stall indefinitely and it's going to take a lot of pressure to make them act. In order to do this right, we have to go against the circumspect advice of our key diplomats — Salem, Qubad, and Livia. They're diplomats, and we have no doubt that they are doing everything they can to help us, but their job is to keep the peace and appease both sides. That's not our job. Our job is to make sure holding Shane and Josh has a cost. Our government hasn't done that; now it's up to us.

A week after going on *Hardtalk* in London, we're back at the White House. This time, when we walk into Dennis Ross's office, a thick tension permeates the room. I find an empty chair directly across from Ross and sit down.

"So," he begins, looking directly at me, "what have you heard from Salem?"

"What have I heard?" I ask, allowing for a long, uncomfortable pause. "We came here to ask you the same question."

No one says anything. Ross and I stare at each other with what feels like open hatred. For almost a minute, nobody moves.

"What is the point of these meetings if you have nothing to bring to the table?" Cindy asks.

"We want you to know that we've secretly passed on a message to Iran that we will hold them responsible for any harm that might come to Josh and Shane," Ross says.

"What does that mean?" Cindy snaps back. "They've already been harmed! Anyway, what consequences? There have never been any consequences."

"Not now," Puneet Talwar responds nervously. "In the future."

"We don't need consequences in the future," Cindy responds with steely resolve. "We need them now."

"You've known for a long time exactly what will get them out," I say. "If Shane and Josh were important to this administration, Shahrazad Gholikhan would be released from prison early, on good behavior. She's close to the end of her sentence anyway — it shouldn't be that hard."

"Our lawyers tell us . . . ," Puneet begins. Cindy suddenly stands up, then Alex and I join her.

"We've heard this before," Cindy says, "and we don't believe it. Unless you have something new to say, we're leaving."

The three of us walk out to the hallway. While we're waiting to retrieve our cell phones, Alex suddenly looks at us sheepishly and says he's changed his mind and will join us outside in a minute. Before we can object, he goes back inside.

I know he's going back to play "good cop" and a part of me doesn't blame him. After all, any one of our intermediaries might be able to get this job done, but only the U.S. government could do it for sure if it wanted to. In the last year, I've gone from feeling like the U.S. government truly regards us as a priority, granting us meetings at the highest levels, to feeling insulted by how manipulative and condescending its representatives are. The United States wants to avoid giving in to Iran at any cost. It doesn't want to show weakness. Ironically, Iran has shown its strength anyway, simply by resisting pressure and holding Shane and Josh this long. All we can do at this point is continue to assert pressure in the media, in hopes that the United States, Iran, and all the other important players will want to wash their hands of this badly enough to act decisively.

A few days later, the Iranian prosecutor Dolatabadi issues a direct response to the media I've been doing. "Sarah Shourd," he says, "has left the path of fairness." Hours later, we receive a frenzied call from the State Department's Iran Desk requesting a family conference call. Both sides feel threatened. We have their attention.

94. JOSH

It's June 4 — my birthday — and I'm trying not to feel sad. I can't help thinking how nice it would be to celebrate with friends and family. Almost every day in April and May, I asked a guard or an interrogator for birthday presents. I told them that I want bed sheets, a pen, a phone call, and a tour of the city. Listening to my requests, Shane always smirks. Then, he usually tacks on items that he thinks are more realistic: a chessboard, a cake — or perhaps pizza. Months of requests have amounted to nothing.

I make one request to Shane: that for my birthday we eat dinner at a table. When our meal arrives, Shane helps me disassemble the bed frame and lay its metal sheet on two stacks of blankets. Shane then covers the metal tabletop with a colorful shawl that my brother sent us. I sit on our plastic chair, and Shane sits on a stack of blankets facing me.

Shane makes the day special. He cooks apples in a plastic bag in a thermos of hot water to make apple pie for dessert. He makes a delicious carrot-cheese concoction. Normally, we chop carrots for ourselves by chewing them up and spitting them out, but today, Shane takes special care to chop them with plastic spoons, breaking several utensils in the process.

This dinner is beautiful — the loveliest thing we've done in a long time. I take the sprouting onion down from the windowsill and place it proudly on our table. We laugh at ourselves for wearing our button-down prisoner shirts — the ones they make us wear to interrogations — as our way of dressing up for the occasion. Shane drives me nuts sometimes, but anybody would in this circumstance. I'm grateful he puts up with me. It is a sweet gift to hear him say, "If I had to be stuck in a cell like this with somebody, it would be you." I look at him, and without hesitating, I tell him the same thing.

95. SHANE

A few days after Josh's birthday, a guard stands at the door of *hava khori* with a bird in his hand. Actually, he has a newspaper in his hand with a bird lying on top of it. This man displays it to us like a boy would show his friends: Look!

The bird is not yet a fledgling and it appears to be barely alive. One little wing moves its body slowly around the paper. Its eyes are still closed. I know its mother. She has been in a state for days, screaming at us whenever we're in *hava khori*, angry that we are too close to the nest, which I imagine sits just up over the wall. This little one must have fallen out of it.

There used to be a little bird inside, just down the stairs from *hava khori*. This one wasn't wild, but a yellow finch. It lived in a cage, hanging from the ceiling next to a security camera, above where the guards sat and read their newspapers. Whenever we came to *hava khori*, I always tilted my head back to look at it. For some reason, the guards never scolded me for doing this. Some of them took delight in it. *"Parandeh?"* they would say sweetly. All day long, it would watch its reflection in the mirror and flit around in the cage, creating a pleasant tinkling noise against the wire frame.

Now, back at our cell, the guard holds the newspaper and the bird forward. *"Mikhay?"* he says. He makes little feeding motions with his free hand, stretching his head up into the air and chopping his beak at his hand like a hungry nestling. "Mother," he says, pointing to Josh and me. He is asking if we want to be the bird's mother. Josh looks bewildered. I say yes.

"Soak bread in water and feed it to him," the guard tells me with authority. I know that a diet of bread kills birds, but I nod in assent.

When he leaves, I place the newspaper on the ground. We stare at it. Its little body is breathing in quick, shallow breaths. Josh still looks bewildered. "Let's make a nest for it," I say. We take the one little metal bowl we have and we fill it with shredded newspaper. I gently put the bird in it. It barely moves.

As a kid in rural Minnesota, I was rescuing fallen baby animals all the time. There were several birds I tried to wean and once, a chipmunk. I would set my alarm every two hours at night, wake, and feed it. But we had syringes then. Baby birds won't eat out of your hand.

I start mashing up some bread and soak it with tahini, figuring that ground-up sesame is the closest we will get to birdseed.

Then the door opens. Food Guy, the one who takes our canteen order, sticks his head in. "Where is the bird?" he asks peremptorily in Farsi. I am sure that word spread and the higher-ups scolded some-one for being foolish enough to give us a pet. I carry the bird-in-a-bowl over to him and make to hand it over. He smiles at the sight of it and hands me a syringe. "For water," he says.

I make a little mixture of water and tahini in a plastic cup and fill up the syringe. The bird won't open its mouth, so I gently pry its beak open and slowly squirt the mixture in. It swallows some and some flows out of its mouth. Suddenly, it comes to life, flapping its wings in the bowl. It does this intermittently for hours.

"If this bird lives, I say we let it go as soon as it can fly," I propose to Josh. He agrees.

Like everything, we have to decide together how to deal with this animal. Our power dynamic is delicate. If I try to assert control over the bird, I'm afraid Josh will take offense and assert control himself. Everything that enters this cell belongs to us equally unless agreed otherwise. But I don't want to hand this little life over to Josh. The only animals he grew up feeding were some fish in a fish tank. I know his intentions are good, but I want this bird to live. Still, I know the worst thing to do is to act like I know better, even when I do.

These kinds of calculations are starting to wear on me. There are so many things building that I want to tell Josh but don't. I want to tell him he has too many vocabulary lists hidden under the carpet. They are forming a lump and someone might see them. I also want to point out that I don't like the way he sloppily puts pieces of writing — all of which are contraband — under his blanket and leaves them there, rather than returning them to their proper hiding place under the carpet. I want to tell him it bothers me the way he sits across from me at lunch and stares down at his food, not saying a word, and that I don't like it that when I wake up in the morning to go to the bathroom, he just lies there staring at his book when I pass, not even saying good morning or acknowledging my existence. And I don't like how much he's been banging on the door lately when the guards don't come quickly enough. What is the hurry? Why work yourself up over a bar of soap or toothpaste?

But I know most of these things are petty, so I don't mention them. I just feel the frustration rise day in and day out, constantly pushing it down in the interest of keeping the peace. Somehow, it feels like this bird has come at the perfect time, giving us another being to focus on, something to care for when we are losing our ability to care for each other.

"Do you want to feed it?" I ask, holding the bird out to him.

He does, but he looks to me for guidance. "Just hold it in your hand and stick the syringe in its mouth. You will have to push it in, but just be careful." He picks it up. "Softer!" I say. He is gripping it too tight. "Keep your grip loose." He loosens his grip and slowly shoots the liquid down its throat. "How was that?" he asks.

"That was great," I say.

We go back to our beds and take up our books. "Shane, I want you to take the lead on this," he says. "Just tell me if I can help." I'm relieved. Maybe I didn't have anything to worry about. Or maybe Josh is being intentionally generous to ease the tension.

I think the bird will make it. It is almost flapping itself out of its nest. I give it a light bath with the corner of a towel, wiping the tahini oil off its chest. I change the old newspaper out of its nest. There are some wet feces in the bowl. A good sign.

I feed it at midnight and go to sleep, waking once before dawn to feed it again. I hear it flapping on and off throughout the night.

When I open my eyes in the morning, the bird is still. I get up to look at it. There are scraps of newspaper outside of the bowl, pushed out of the nest by all its movement. In the bowl, half-covered in newspaper strips, the bird is stiff.

We wrap it in newspaper. To its little coffin, Josh adds a plastic bottle cap full of halva. It is our favorite treat, sweet and made of sesame. We set the bundle gently in our trash can.

A couple days later, Food Guy comes back. "Where is the bird?" he asks in Farsi.

"Dead," I tell him.

"Dead? Where is it? What did you do with it?"

"We put it in the trash."

"In the trash! You just threw it away like garbage?" He looks at us like we are uncouth and insensitive.

I look around us, making sure he sees me scan the cement that encapsulates us. Then I look at him, amazed. What did he expect us to do with it?

96. JOSH

It's ten days after my birthday. The black asphalt is hot at high noon. I'm in the short-sleeve collared shirt that I wore while hiking in Iraqi Kurdistan twenty-two and a half months ago. Instead of the dusty jeans that I also wore on my last day of freedom, I now wear tight Euro-style jeans given to me last year by the Swiss ambassador. It's my first time wearing "real" clothes in a long while. A van pulls up, and Shane and I climb in.

We curve around the Evin Prison compound, winding past an eight-story prison hospital and the administrative building where we once met the Swiss ambassador.

Dumb Guy sits in the passenger seat and Father Guy sits in the back row. A soldier stops the van. Another checks the driver's paperwork. Another does a routine check of the underside of the car. Then, the large, white steel gate — that last barricade between Evin Prison and the civilian world — opens wide.

The interrogators are taking us to eat pizza in a park — a belated treat for my birthday. Before entering the stream of street traffic, the van accelerates directly toward a gaggle of civilians on the road's edge. The driver slams on the brakes in front of the crowd, and the side door slides open. A man enters carrying a professional video camera. As the door closes and the van careens into the flow of traffic, I hear Shane's voice from behind me. He speaks quietly, just above a whisper. "Josh, do you recognize this guy?"

"Who?"

"The cameraman."

"No."

"You don't recognize him at all?"

"No."

"I think he was filming at court."

Shane thinks the cameraman works for a TV station that will air our trip to the park. Dumb Guy promises it won't be televised. Shane wants to head back to Evin so this event is not used for propaganda, but he leaves the final decision up to me. Dumb Guy hears us discussing our options, and he interrupts from the passenger seat. "This is your last chance. I worked hard to make this happen," he says. "I promise this will not be on TV. If it is on TV, you will see it. You will know if I am lying, and you will not trust me in the future."

Dumb Guy doesn't want us to turn around. He's arguing with us but giving us the choice. He never gives us choices like this. He also never puts his trust on the line like this. Oddly, I believe him.

And I just feel like going outside. Can't that be a good enough reason? Yet I know, as with everything, I need a thorough rationale for every decision I make so that I can explain myself to Shane and to the interrogators. My head spins with more reasons to continue the trip. Nelson Mandela toured with his captors at the end of his imprisonment. Why shouldn't I? Maybe this is the end of ours. Why should I refuse better treatment just so they cannot say that they treat me better? I would never turn down books to show they deprive me of literature. I would never refuse a roommate in order to proclaim that they keep me in solitary.

I open the window wider to help me decide. The air rushes onto my face. I see images that remind me of a past life: yellow taxis honking and swerving around each other, huge flowering trees lining the streets, fruit vendors on the sidewalk, and businessmen walking past high-rise buildings. There are children playing in a playground, old people sitting and reading the newspaper.

There is no way to know whether Dumb Guy is lying or not, whether our experience will be aired on the nightly news or not. It's a question of faith, but normally — in the cell — everything is calculated. Shane and I chart our exercises, count our books, measure our storage space, and mark the days on the wall. Whether to hunger-strike, to write a letter to the interrogators, or to ask for Iranian

DVDs instead of films from Hollywood requires a rational, mental debate of the pros and cons.

I feel my heart stir. Then I do something I feared I had forgotten how to do: I make my decision based on intuition — a trusting intuition. "We want to go." I project my voice to make sure Dumb Guy hears me in the van's passenger seat. "We don't want to turn around."

I stick my head far out the window. The unconfined air streams around my face and neck. I close my eyes and enjoy the sensation of movement. The air currents against my eyelids slow down as the car decelerates. I sense the warm summery stillness and open my eyes to a traffic jam.

Now would be the perfect time to run. We're already wearing street clothes and sneakers. This thought makes me feel even freer.

Eventually, we drive partway up the giant mountain overlooking Tehran. We pull over by a park and cross the street to enter.

Subtropical trees arch overhead; junipers, tulips, and roses beckon from beyond the trail. Thick, moist air fills my lungs. Dragonflies, squirrels, and bees welcome me. My shoulders relax into my back and my knees bend more than normal as I pace through the canopied walkway. The uneven terrain underfoot contrasts to the flat hallways of Section 209, and I sense the earth's complexity through my shoes. I look left and look right into the lush landscape, but my body keeps walking forward, unaccustomed to lateral movement.

My mind wanders. I imagine hiking with my brother and friends near my old home in Oregon. A warm, dank gust of air fills my nose with the sweet smell of flowers. I look over at Shane, who is mesmerized by the trees. Our eyes meet. He seems to read my mind and nods in agreement. I can't believe I've forgotten what this feels like. Freedom seems magical, far beyond words. I can barely express to Shane how much this contrasts with the cell and *hava khori*.

Yet, the guards and interrogators trail behind us and the cameraman jogs ahead. He signals for us to look forward. I don't submit to his request, nor do I rebel and avert my eyes. I'm committed to making him as irrelevant to my experience as possible. It's my way of practicing being free.

The cameraman is backpedaling and filming. The forested path widens into the open sky and plaza. There are some benches nearby, a kiosk in the distance, and a pond right behind the cameraman. *Splash!* He falls backward directly into the pool of water. The guards run to help.

The cameraman hobbles over to a bench with his arms around the shoulders of two guards. Blood slowly seeps down his leg. One guard cleans his wound; another carries the camera awkwardly. I turn away and suppress a smile. He's done for the day. We no longer have to worry about footage being used as propaganda. I look at the trees and silently thank them, as if, somehow, their spirits contrived the cameraman's fall.

We're told to order ice cream at the kiosk. I remember they gave Sarah ice cream in the days before they released her. I want pistachio ice cream, but they have only chocolate and vanilla. There are a couple young men watching us as we eat. "Welcome," one of them says in English. Do they recognize us from television or is this Middle Eastern hospitality? *"Man zendooniam. Amreekaaii."* I'm a prisoner. American. I speak to the civilian in the same way I'd whisper to a neighboring prisoner down the hall. Likewise, a guard silences me and stands between us.

After devouring the ice cream, I ask for permission to hike. The guards and interrogators have trained us to ask for permission for nearly everything. The guard waves his arm. "Go ahead," he answers, surprised that I asked. We spring from our seats and trek uphill. The guards lazily lag behind. We arrive at another pond. Around its edge, I see clusters of brightly clothed people of all ages sitting and lying on picnic blankets, smoking hookahs, eating lunch, and just relaxing.

We ramble uphill and gaze down at the guards huffing and puffing some distance below. The landscape changes dramatically at this higher elevation — there are no more trees. Shane and I take in a bird's-eye view of the city. High rises and highways mark the landscape. Glass buildings and domed mosques reflect the sunlight to our aerie. I stare silently at the stunning panorama. The world has never felt so vast.

When Father Guy arrives, I ask him where the prison is. He points beyond a protruding ridge and says we can't see it from here. This view is beautiful, but Shane and I don't want to linger. We want to run.

This time we don't ask permission nor do we calculate how much time we have left. Shane and I follow our momentum down a narrow trail. The hillside steepens downward and I zigzag through the trees. I abandon the trail; we're no longer confined to the linearity of prison hallways. I speed up and Shane follows. The guards must be behind us, but they don't feel near. I pick up the pace even more, and the cool, moist air brushes against my sun-warmed skin. This is the meaning of *hava khori*, literally *eating air:* the free world is my sustenance. I feel alive and fresh and fully present. I feel at one with the magical trees with their rustling leaves cheering me on. Suddenly, I hear a guard's voice. He approaches in his muddied dress shoes with beads of sweat dripping from his forehead. He tells us to slow down and follow him back up the hill.

He takes us to the restaurant. Back with the interrogators and guards, I feel the constraints of detention re-envelop us. Shane and I sit alone at a table and open a menu. I order a fresh-squeezed carrot juice, French fries, and vegetarian pizza. The guards and interrogators, sitting at the neighboring table, order a round of Coca-Colas.

"Coca-Cola?" I say to them. "Aren't you guys anti-American? That's Americana at its worst — it's bad for your health and it's addictive. Coca-Cola is the epitome of imperialism!" They are much more intrigued by my carrot juice. Their whole table laughs at me.

Dumb Guy explains the laughter to me with a Farsi word, "*Parandeh. A bird.* You are like a *parandeh* for drinking carrot juice with vegetable pizza."

I also eat fries with spicy salt, and I laugh with Shane as he pours ketchup on his pizza. We savor our food, clinging to these tastes as if they were the taste of freedom itself. When we're done eating, we walk down the path to where we entered the park. The bushes that beckoned us upon arrival continue to call my senses. I notice their flowers in full bloom as I approach the van.

Nearing the prison gates, I prepare to put on my blindfold. My mind returns to marking time. It's 4:30 p.m. It is June 14. We've been gone for four hours. I remember that we've not received letters for almost a month. In the changing room I relinquish my jeans and collared shirt. I brace for the sadness of the cell. All I can think of is receiving letters. I call out to Dumb Guy. "You know in four days you will owe us letters. We'll have to hunger-strike if you don't come back with them in four days. Also, it is Shane's birthday in a month."

"Really?" Dumb Guy retorts. "Then this trip was for Shane's birthday too. I just did you a big favor. You should appreciate it!" He hands us off to a guard who escorts us inside.

I walk down the familiar hallways to my cell with a broader chest, a more open heart, and a more relaxed gait. I inch my blindfold up to my forehead, trying to act like the prison's rules are irrelevant. No guards pester me, except one, who meets me in my cell to point and laugh, saying, *"Parandeh."*

97. SHANE

The prison doctor says I'm anemic. He took my blood the other day. I've lost about six pounds this month. When I run our little loops at *hava khori* lately, I finish exhausted.

My body is falling apart. Sometimes, it feels like there is an animal in my guts running all around. Other times, it's as if there were a lead weight in my belly. I almost always feel like there is something poisonous in there.

Lately, I've been sleeping eleven hours a night. Josh lies in bed for the same amount of time, but he spends much of it trying to fight himself into sleep. I sleep deeply in a world of dreams where I find myself free, and having to quickly return to prison. When I wake, I am glad to know I have skipped the morning and that it is almost time to eat lunch. Josh and I have become more slothful. When we first celled together, we would wake up around six, exercise, and have breakfast before going out to see Sarah at eight. Now, we spend most of the day in our own beds.

I have been on various pill cocktails over the last year, none of which have worked, so one day they take me to a hospital in the middle of Tehran for more tests. Inside, I stand in a bathroom and remove my clothes. I am supposed to give a stool sample in a little cup, but I just stand in front of the mirror, naked, and stare. The person I see is me, but different. My ribs are showing. My head looks larger than normal, and my arms are spindly. There is a vacancy in my eyes. Actually, *vacancy* is not quite right — it's more a look of circumspection, as though the person behind the glass were looking at a former friend he no longer knows how to relate to. It's not just my body that's withering away, but something deeper. It's not an anguish I see so much as a dimming light. I don't feel much anymore. Everything is just a dull, unhappy drone. The free parts of me are fading. This does not sadden me, really. It is just a fact and, like most everything else lately, I just observe it as it floats past my consciousness.

Being out of the cell helps me see the person who lives in the cell, sort of like the way going on vacation helps you see your domestic life from a fresh perspective. I see that for many months now, I have just been rolling along. I don't mind taking orders from the guards anymore. In fact, once I gave in to the prisoner-guard dynamic — I don't know when it happened exactly — things got easier, more pleasant. There is something comforting about it. It's a depressive comfort, sure, but once you look at it a certain way, it's not so bad not having to make decisions. It's not like I acquiesce to everything. We have our hunger-strike game over the letters, but that has become routine. We committed to striking whenever they neglect to give us letters for thirty days, but usually, whenever we near the mark, I'd rather just let the letters slide than hunger-strike again. They will come eventually. And why stir things up?

I squint at myself in the mirror, looking deeper. Something is wrong with me, I realize. I have been letting my life slip away. These ideas of acceptance, this Buddhist seduction, it's all bullshit. The spring air, books by Eckhart Tolle, Ram Das, and Pema Chödrön — all these have been singing me into a stupor. I thought it was spiritual — and maybe it was — but I don't want it anymore. I meditate daily with the goal of staying present in every moment. Sarah almost became a Muslim, and now I'm becoming some sort of quasi Buddhist. It's the same trap. I've been feeling at home — at *home* — here. I have been getting closer and closer to figuring out the big secret — how to be genuinely content anywhere, how to live in the eternal moment. But all this is just the flip side of my old obsession with standing up to the guards and interrogators. Both are just ways to trick myself into feeling free inside prison. But I'm not free.

I am just becoming institutionalized and cloaking it in spirituality. Thinking about the outside — about freedom — just makes me suffer, so I have locked those thoughts away. I don't even think about Sarah much anymore, except when I get letters from her. Josh and I have kept up the commitment we made to her — to have a moment of silence and think of her every day during the evening call to prayer — but even that has become rote. I am becoming less and less able to pull her image up in my mind. My thoughts wander. We have pictures of her all over our walls, but it's hard for me to bring even those to life.

I shit in the little cup, dress, and rejoin the guards in the waiting area. There are couples everywhere. A long-haired old man and a short, hunched woman approach the reception desk, holding hands. A toddler waddles across the room, her parents trailing behind. I don't know the last time I've heard a din of conversation like this.

Josh and I need to get out. I know if I wait any longer, I will give in. I think we both will. We have already gone through the first four stages of grief: disbelief, yearning, anger, depression. Now we are a good way into acceptance. We need to do something before the outside becomes a total fantasy. This month, with our trip to the park and this trip to the hospital, I have seen more trees, more women, and more babies than I have in over a year and a half. We can't keep missing the outside world.

In the coming weeks, I take more trips to the hospital and other clinics to take tests for my anemia. I record little details in my mind. I come to know that behind the hospital there is a warren of residential housing. I know that two perpendicular streets alongside the hospital are big and busy and usually clogged with cars, so if someone were to run, it would be hard to chase him down by car, especially if he were running against traffic. When they take me out, they usually like to put me in civilian clothes and tennis shoes. They give me real laces to replace the strip of T-shirt fabric we usually tie them with. They always take me in a regular little car with clear windows. I am never in restraints. The guards are usually different each time, but they tend to be old or overweight, or both. And they are almost always wearing pointy dress shoes. I know I could outrun them, even barefoot. I am young, I exercise daily, and my freedom is at stake.

For weeks, I ask myself if I am really ready to do it. I decide that

I am. I can't go through another winter here. All I need to do is convince Josh.

We used to talk about escape, but our plans were just stories, always of the jailbreak sort. We'd punch out a guard and lock him in our cell, then let all the prisoners out and make a dash for it. Or we'd open our cell door by reaching through the little window—which we used to be able to do in our old cell—and creep down the stairs at night and get into the trunk of someone's car. Then we'd wait for the morning.

One day over dinner, picking at my plate of potato salad, I gather the courage to say to Josh, "I think we should escape." He looks at me like he isn't sure where this is going, but he is curious. "I think I know how we could do it," I say.

He leans back and sets his eyes in a way that lets me know he has my full attention. "Let's hear it," he says.

I explain that one of us would go to the prison doctor and complain of intestinal problems. I reason that I should probably do that, since I'm the one who's been having issues. They would put me on a medication like they always do; then after two weeks or so, I'd go back and say the problem was getting worse. They would put me on another medication. This would set the stage, so they would feel like I have a problem they can't figure out. Shortly after that, Josh would go in and complain of the same symptoms. Over the period of a week, he would complain of it getting worse. Eventually, they would take us to the hospital for more tests.

Josh is absorbed by my scenario. I pull the pen out from under the carpet, grab one of our old letters from home, and draw a map of the area around the hospital. "Every time I've gone, they've parked on this street," I say. "I have walked fifteen to twenty feet away from them without a fuss. Their guard is always down for some reason. We would get out of the car, walk leisurely ahead, then book it around the corner. At least one of them would chase after us and another would probably get in the car and get on the phone right away—"

"Don't you think they would shoot us?" Josh asks.

"They don't have guns," I say. He looks at me skeptically.

"No," I say. "I am certain. I've looked closely at their waistlines and pockets. They don't carry guns to the hospital. For some reason they try to keep a low profile . . . So, we run against traffic and outrun whoever is chasing us. We just run until we lose them and we hide out in a Dumpster or a rooftop till nighttime.

"Then we start to walk to Azerbaijan. No one will expect it. We walk only by night, following roads at a distance. If we stay near the roads, we'll be able to see road signs that can tell us where we are going. We find spots to hide and sleep during the day. The problem is we won't have money and we won't be able to talk to people, so we'll have to steal food too."

"Wow," Josh says. "They would be looking for us everywhere. This is pretty crazy." He pauses. "But I am interested. I'm obviously going to need to think about this for a bit."

In the following weeks, our runs at *hava khori* are more intense. We run faster and I don't let myself skip it if I am tired or lazy. As I turn those circles, the walls become buildings. I scan them for doorways to duck into. I feel a guard always on my tail.

Josh and I refine the plan and turn over different options. Should we steal a car? Maybe. Could we jump someone for a cell phone, call our family, and have them wire us a ton of money we could use to bribe someone to smuggle us over the border? Probably not. We piece together a map of Iran in our minds from the short intervals that maps of the country appear on the news. We judge that if we walked three to four miles per hour, we could do at least thirty to forty miles a day, getting us to Azerbaijan in about two weeks.

One day, we see a topo map on TV and my heart sinks. The region north of Tehran is full of mountains. I had an image in my mind of us walking through fields with the road in sight, but if it is mountainous, we'd probably have to walk *on* the road much of the time. We'd get caught for sure.

"What about the Swiss embassy?" I ask Josh days later. "We could get in a taxi, go straight there, and get inside even before the word got out. The Swiss would have to take us in. And they couldn't turn us back over to the Iranians. It would surely break some international law. We are being held without trial here. We would be asylum seekers."

As the days and weeks pass, Josh gets more tense. He talks less and looks depressed. He always seems to be brooding, looking blankly at the floor when he isn't reading. He breaks the silence only when he thinks of new contingencies we could come up against. How will we get water? How little can we eat in a day? What if it *does* snow? What will happen if we get caught? Is Azerbaijan even the best destination? What about Turkmenistan?

I dream nightly about escape. I find myself in the empty streets of Tehran in the middle of the night. I am on the lam in California, hiding out in beachside hotels. I walk with Josh through long plains of knee-high grass.

One day, at the end of June, we see a ticker: "Americans to Go to Trial July 31." We postpone our decision for two weeks, and we wait to see what happens in court.

98. JOSH

That ticker makes it feel real. The momentum seems to be building on the campaign: in May a phone call, in June the trip to the park, in July our "final" court date. Perhaps the most universally revered living Muslim, Muhammad Ali, amped up his support. I'm feeling secretly confident. Ramadan and the UN General Assembly — which coincided with Sarah's release — soon approach. Alex keeps writing a new Dylan quote in his letters, "Freedom just around the corner for you."

Yet, with all this momentum, the most embarrassing thought springs to mind: I want to stay in prison longer. There is more that I want to read. And I want to finish the novel I've started writing. The protagonist struggles to admit that he cannot find freedom behind bars, no matter how hard he tries, and eventually he decides to escape.

Oddly, the desire to stay in prison used to menace me when I was most desperate, in solitary confinement. In solitary, I wanted to prolong my imprisonment in order to memorize an expanded multiplication table, or to work my way up to one hundred squats and fifty pushups in a row, or to perfect the Morse code, or to juggle dried oranges fifteen hundred times in a row. Back then, the thought was functional. It validated my daily activities, making them feel worthwhile. Now, in much better conditions, this temptation to stay in prison resurfaces.

Anyway, what do I have to look forward to? My family probably resents me for putting them through this. I still dream of my friend Jenny, but Shane thinks I'm delusional for focusing on her. He's probably right. Mom writes that Jenny's on the rolling hunger strike, but for all I know, she may be married with children. In prison, at least I have a good excuse for feeling crappy.

Alex's Dylan quote loops in my head again and again: "Freedom just around the corner for you." He never includes the next line: "But . . . what good would it do?"

99. SHANE

Josh grabs a bottle from the garbage can on the way to *hava khori*. It is one of those green plastic ones — a one-liter. As soon as I see him grab it, I tense up inside. I know he wants to add it to our collection. We have seventeen plastic bottles in the cell right now, all of them filled with water. Six of them are stacked in a pyramid under the bed and eleven are standing under the sink in our tiny bathroom. Then there are four or five paper juice bottles, each also filled with a liter of liquid. And we have about eight little quarter-liter bottles of dough, the salty yogurt drink. We use all of these as weights, putting them in plastic bags and lifting them. But there are never enough. When one person is doing the bench press — lying on a stack of blankets and benching a bar we broke off the bed frame with bags of bottles hanging from each side — there aren't enough bottles for the other person to do tricep extensions at the same time. Josh doesn't want his exercise routine to be dictated by mine, nor do I want mine to be determined by his. So we keep amassing bottles. But today, I feel like we have too many.

We used to have all seventeen bottles under the bed, but a few weeks ago I proposed moving them to the bathroom. We could put them under the sink and along the wall, I said. It would be easier than reaching under the bed to stack them every day after exercise and it would clear up some of that space. Josh rejected my idea initially, then said he would agree to it if we kept six bottles under the bed. Josh doesn't like to feel crowded by bottles when he is standing in front of the sink or squatting over the toilet. Everything has to be a compromise. Nothing is easy anymore. Where is he going to want to put this new bottle?

When we get to *hava khori,* he doesn't say anything about the bottle, just tosses it in the corner with the blankets we brought out to stretch on. I don't say anything either, just start stretching my legs to start our running routine.

These moments, like the majority of our time together, tend to be silent lately. Ever since we found out that a new trial date is coming up, they have been quieter than ever. When we learned about the trial, we acknowledged to each other that things usually get difficult in here when some significant date is pending. We committed to being vigilant about supporting each other, easygoing. But it isn't always so easy.

We each start walking the perimeter of the courtyard at opposite ends. After a few laps, I say, "Ready?" and we jog. I count thirty-five rounds, then yell, "Switch!" We turn around, Josh counts thirty-five more, and we switch again. We do this six times, which, according to our measurements, is a little over three miles. One lap is eighty-five shoe lengths around and a shoe is eleven inches. We know this because we measured a shoe against a letter sent from home and we know a piece of printer paper is eleven inches long. As we run, I can't stop thinking about Josh and his bottle. Is he going to ask me about it?

When we get back to the cell, he holds the bottle up casually. "Got another bottle," he says.

"I think we have enough," I say too quickly. I'm trying to be casual about this, not give away the fact that I am overly concerned, but I just blew it. I know he sees what just happened. He sees everything.

He looks at me with that look of feigned confusion that drives me crazy. "Enough?" he says. "But I thought you wanted more so we could exercise at the same time?"

"Where do you want to put it?" I ask him.

"Oh, I don't know," he says as if considering it for the first time. "Under the bed?"

"I don't want more stuff under the bed," I say, again a little more insistently than I'd like. "It's too crammed. I'd be okay with keeping it if we put it in the bathroom."

He goes into the bathroom as if to mull it over and comes back out thirty seconds later. Now he has a face I know well, one he cannot control. His lips are pressed together and he is frowning. His jaw is clenching slightly. His eyes are downcast and glassy. This is the look Josh has when he believes he is hiding his anger. "I don't want to put it in the bathroom," he says blankly. We stare at each other.

"Well," I say, letting a pause linger, "I don't want it under the bed. I don't even think we should have another bottle. But I'll compromise and take another one if you agree to put it in the bathroom." Checkmate. Actually, I don't really like having the bottles in the bathroom either. The array of them has become a sanctuary for cockroaches. But I will never admit this to Josh. It would only strengthen his position.

He is silent, still staring at me. "There is no room under the bed," I say. "And actually, I've been noticing that your stuff under the bed has been sliding over to my side." His face flashes something intense, as if I used a weapon he himself was about to unleash.

"Well, I've been noticing that I have a lot of common stuff over on my side!" he says.

I get on my hands and knees and look under the bed. "I don't see it," I say, "but if you look at the center bar here, your stuff is over the line."

"You know that bar slides around," Josh counters. He pulls the blankets off the bed, throws them aside, and lifts off the mattress. We both pull away the metal plates that hold up the mattress, revealing a rectangular section of floor packed with stuff. There are about sixty books stacked along the back, all sent from home. Since so many have been coming in lately, we usually give books we are finished with to Ehsan to donate to the prison library, in the hope that the next English-speaking prisoner will have something to read. Most of these books under the bed are those only one of us has read. We each tend to lose interest in a book once the other has read it. It's all part of the little ego game we play. If one of us shows interest in something, the other disparages it subtly, often by not deigning to read it. Our reading choices are one of the only ways we have of distinguishing ourselves. In other ways, we are becoming one person, and I hate it. The only intellectual input we have is books, so I read different books than he does.

We aren't even planning to read all these books under the bed. It's just that we each have different ones we are not willing to give up. For me they are the dense works of philosophers like Heidegger and Merleau-Ponty. I don't intend to read these — I've tried and given up several times — but I want to keep them in case the censors cut off our book supply for some reason. If we run out of books, it is good to have something to grapple with, something that will take a long time to understand. Then there are the books that are our favorites. Of the two hundred fifty or so we've read, these are ones we might reread in a year or two if we are still here, books like *The Grapes of Wrath, The Brothers Karamazov, War and Peace,* and *The Waves.* And there are also the biographies we want to hang on to for reference, such as Martin Luther King Jr., Stalin, Mao, Gandhi. We never look at these books after reading them, but we keep them anyway. They provide some sense of security. The ones we actually do reference — mostly poetry or history — are spread around the cell on our food-tray bookshelves.

There are almost one hundred fifty volumes in the cell. Ridiculous.

As Josh lifts the mattress off the bed frame, he points out that there is one more stack of books on his side than there is on mine. He also has our extra bottle of ketchup and the antenna we've been stashing in case they ever decide to disconnect the cable from our TV. I counter that I have the cardboard chessboard we never use and the tahini container filled with chess pieces made out of tinfoil yogurt tops. I also have the extra bags of spices, spillover from the spice rack — a "shelf" made of plastic spoons stuck into the grating of the heater side by side in a row, holding up quarter-liter bottles filled with salt, pepper, sugar, cardamom pods, ginger, curry, cumin, and cinnamon.

Most food is now divided into private property, since we save and gorge on different items at different rates. The longer we've been celled together, the less we've shared. The items now divided between our two sides under the bed include two fresh boxes of dates; about three hundred letters from home; four containers of Jif peanut butter — two regular and two with honey; three tahini containers; a kilo each of walnuts and almonds; two bars of chocolate;

four little squeeze bottles of honey; two plastic containers of halva; and forty packets of digestive biscuits, each containing twenty biscuits. Then there are our separate stacks of clothes and two pairs of long johns and two sweatshirts rolled up and stuffed in the back for wintertime.

Somehow, as we look over all of this stuff, we both feel like we are being cheated by the other. The only solution, we decide, is to take everything out that we share and divide it up. We forge ahead, forgetting that not too long ago, we didn't care about these square inches of space. We used to be generous, each offering to take on little burdens just to relieve the other. Now, we are tearing the cell up over a water bottle.

From under the bed, we take out the box filled with about forty of the little jams and honey we get for breakfast. We pull out the two boxes of cookies that look unopened—they are still carefully wrapped in cellophane—but in fact contain our secret prison diaries. There are the two boxes of fermenting dates we hope someday to make pruno with, a little bag containing five bars of halva, another bag of year-old pistachios with shells that can't be opened, and a little bag of ancient candies that Sarah kept from our first month in Section 209 and passed on to us before she left. There are the juice bottles, our bread bag, and a little box of GRE vocabulary flashcards. We combine these items in several little piles that seem about equal in size; then we take turns picking one pile after the next. Josh measures the length of the bed to find the exact middle and we scratch a mark there into the paint.

None of this has an impact on the decision over what we will do with the new water bottle. Neither of us budges on where we think it should go, so we throw it away. If we ever find any one-and-a-half-liter bottles, we decide, we will trade those out for our one-liter bottles under the sink. That is our compromise.

Things have never been this bad between us. After we put our cell back together, we retreat silently to our respective sides of the cell. We don't acknowledge that we have both been wracking our brains with escape plots and court scenarios for weeks. We only acknowledge the bottle.

100. SARAH

It's July 30, 2011. I've come to notice how much the Iranian government loves to choose these special dates as a passive-aggressive way of sending messages. The day I was meant to be pardoned was September 11. Shane and Josh's first trial session was the anniversary of Iran's revolution. In Iran it's already July 31, the two-year anniversary of the day we were captured in 2009. Today, a representative of the Revolutionary Court has stated, the second and final session of Shane and Josh's trial will be completed. Then, a decision will be made.

The Islamic Republic of Iran is not allowed an embassy in the United States. They do have a small office in a high rise in downtown New York where their UN ambassador, Mohammad Khazaee, presides. Yesterday, we held a protest at the base of this building. More than a hundred people showed up, bearing signs and banners that read TIME FOR COMPASSION and SHAME ON YOU, IRAN. A row of news cameras flanked the stage as family members, Muslim leaders, and human rights speakers took turns addressing their grievances, hoping to make the men sitting behind tinted glass, thirty-four stories above us, a little uncomfortable.

The protest was the culmination of a week of action dubbed "Two Years Is Too Long!" leading up to the trial. Yusuf Islam (formerly known as Cat Stevens) published an open letter to religious leaders in Iran asking for a compassionate, Ramadan release. Our friends in the San Francisco Bay Area organized a call-in blitz they called "A Million Voices," mobilizing thousands of callers to jam the phone lines of the Iranian Interests Section's pseudo-consulate in Washington, DC, every day of the week. Noam Chomsky released one of his many statements in which he called on the Iranian government to show that it makes a distinction between the American government's policies and its people. Tom Morello, formerly of the rock band Rage Against the Machine, posted a YouTube video of himself calling in. "Hi. I'm a supporter of Shane Bauer and Josh Fattal. I urge you to release them immediately . . ." By the fifth day of "A Million Voices," the operators at the Interests Section began to beg people to stop calling.

"We're sorry," the receptionist told the callers, "but there's nothing we can do about this."

Over the last month, we've helped raise a storm. All of our intermediaries have gotten hopeful, if inconclusive, indications. Now, all we can do is wait.

After the protest, my mom, a handful of friends, and I take the train to a small town an hour north of New York City. The house we've been lent is like something from a Victorian novel. It stretches out room after room, seemingly without end. At first, the four of us spread out, occupying different wings, but by midnight we're all back in the kitchen, working on our computers, drinking tea, and looking out the tall windows at our own reflections framed by the blackness outside. A part of me knows that we won't hear anything tonight, but none of us wants to sleep. As the sun rises over empty streets, I wonder if that really was my last, long sleepless night. It's the first day of Ramadan.

101. SHANE

It's July 31, exactly two years since we were captured, and we are back at court. It has been almost six months since our last hearing. This time, they keep us in our prison slacks, rather than have us change into street clothes. There are no cameras in the courtroom. According to the news ticker we saw weeks ago, this will be our last hearing. Josh and I have decided to answer all of their questions as briefly as possible. We want to make sure they don't drag this out any longer. We just want our guilty verdict and our sentence today.

Josh has our secret note in his pocket. This one contains an extensive key to a code language our family can use to tell us things they have been afraid to say in letters. According to the code, if they refer to a fictitious name, we will understand the first initial of the first name to represent a continent and the first initial of the last name to represent a country on that continent. Sam Victor would mean "Venezuela" and Adam Sawyer, Syria. And if the two went clothes shopping together, that meant they were negotiating. Sporting events refer to nuclear talks. William Shakespeare means freedom. The note also contains a list of sixty book requests. I want Middle East history. Josh wants economics. If we are going to be stuck here, we want to study.

The judge enters and opens court. The same interpreter from the last session translates for us. This time, the judge doesn't ask to see Sarah — he says she will be tried at a "later date."

The judge says the prosecutor has more questions for us, but our lawyer is first allowed to stand and present his opening statement.

"Your Honor," Shafii says, "tomorrow marks the first day of Ramadan, the holy month of compassion and forgiveness. There is no evidence that my clients are spies. In fact there is plenty of evidence that they are not. On the issue of the border, I cannot say definitively whether they crossed or not. But I can say this: if they strayed across our border and entered our great country, it was not intentional. This is a small transgression. In the spirit of this most holy of months, I ask you to pardon them."

It feels like a helpless argument from a powerless man. But he is right. The timing *is* auspicious. Today is the two-year anniversary of our imprisonment. Josh and I have become almost certain that they are rushing to give us a guilty sentence so as to release us during Ramadan and boast about Islamic compassion. We are both hoping that when we leave here today, we can forget about escaping.

The prosecutor stands up behind his podium and responds to Shafii. "When we last held court, these two said they were not spies, but they should not forget that there is *evidence.* The court has *evidence!* Think for a minute," he continues. "If these people were not important, if they were not spies, why would the United States government be trying to help them?" *What is he talking about? Does he know about some kind of "help" that we don't?* "These people say they are journalists, but this is typical. It is nothing more than a disguise. They collect pictures and write reports of the region to send to their government!" He is starting to whip himself into a frenzy. "They use the disguise as reporters to get close to Palestinian communities and Hamas to gather information!

"When they were stopped by our border police, they were on a mission given to them by the CIA at the U.S. embassy in Baghdad to meet with PJAK," he says, his face now red and his hand constantly pumping into the air, as if he has tapped into some reservoir of national rage. I see where this is going. For the first six months of our detainment, we always thought we would be linked somehow to the Green Movement, but for the last couple weeks, the government has been fighting Kurdish guerillas — PJAK — in the country's west. So they connect our case to this — the issue of the moment.

"And Joshua Fattal! Mr. Fattal is a Jew from Israel! Under the guise of students and journalists, these two carry out missions around the region for the intelligence of the U.S. and Israel. These two deserve the maximum penalty for threatening the Islamic Republic of Iran."

He sits. I feel a pinch of fear. I bury it. *This is a show, remember. Stay cool.*

The questions begin. All of them come from the judge, not the prosecutor.

Q: After you were arrested, the FBI instructed your families not to speak to the media. Why did they do this?

This is idiotic.

A: I was in prison after I was arrested. There is no way for me to know what did or did not happen, or why.

Q: You have a counterfeit passport that you have used to travel to Syria, Yemen, Iraq, Lebanon, Sudan, and Israel.

A: I don't have a counterfeit passport, but I do have a second passport. This is so I can apply for visas, which sometimes takes a long time, and still be able to travel. This is common for journalists working abroad. Any U.S. citizen can do it. And I didn't use it to travel to most of the countries you mentioned. You can look in it yourself to see what stamps are in it.

The prosecutor cuts in. "You see the way he answers these questions? His methods show clear CIA training! He is denying things we have *evidence* of. There are other indications of CIA training too — in their interrogations, their answers were nearly identical." *What?* I look at Josh sharply. In the last hearing, the prosecutor said the opposite: that our answers were inconsistent, which showed that our stories were spy covers. Are both consistency and inconsistency evidence of espionage? Josh is rubbing his forehead.

The judge hands over a piece of paper and tells me to write my response.

A: Our answers were so similar because we were telling the truth.

It is hard for me to understand the meaning of all this. I get why a government would go through the motions of a trial even when the outcome is decided in advance. They're called show trials. But why have a trial when you don't *show* it to the public? They don't even have any cameras here. Maybe my thinking about the purpose of our trial has been wrong all along. Maybe we are on trial just because everyone needs to believe they are acting within the law. Politicians, judges, and prison guards don't want to feel like rogues.

They turn to Josh and question him on being Jewish. While he explains that he's visited his family only a handful of times and doesn't even speak Hebrew, the prosecutor cuts in. "You lived in Israel until you were two years old!" he shouts.

Josh throws his hands up as if to say, "What's the fucking point? You want me to argue where I was *born?*"

"Stay cool," I whisper.

"I was born in Boston," Josh repeats. "I have only visited Israel a few times."

Shafii objects. "What is this?" he asks the judge. "What does his religion have to do with this? Even if his father *is* from Israel, even if *Josh* was born in Israel, which he was *not,* that would not be evidence of espionage."

The prosecutor retorts, "How can you say this is irrelevant? Israel is our greatest enemy! His connection to Israel is something we cannot ignore!"

The judge turns his attention back to me:

Q: You receive money from America and Israel to travel around the Middle East as a student and set up operations in different countries. You establish yourself in Kurdistan because it is a good place to attack Iran from. You learn Arabic because it helps you establish yourself in conflict areas. You carry a sophisticated camera.

I am getting exasperated, but I stay composed. "I don't understand what you are asking me," I say.

The judge holds his hands out and smoothes them over an invisible globe as he says, "Basically, you are at the forefront of an American-Israeli conspiracy against Iran. Please respond."

A: I have nothing to do with the American or Israeli governments.

I carry a camera because I am a photojournalist. We never "established" ourselves in Kurdistan. We were only there for three days. I haven't been a student in years.

Q: How many times have you been to Iraq and how did you get a visa?

A: I have been to Iraq twice. The first time I got a visa from Baghdad and the second time I got a visa from the Kurdish authorities on the border.

The prosecutor jumps up again. "See! Now Mr. Bauer is changing his story! Now he says he *has* been to Baghdad."

"What?! I have always said I have been to Baghdad," I say. My heart is starting to race. "In the last hearing, I talked extensively about my trip to Baghdad. I answered several of your questions about it and spoke about it in front of the camera. I will repeat what I said then: I went to Baghdad as a journalist five months before we were captured—"

"*Noooow* you say you were in Baghdad," he retorts. "Now you *admit* it!"

"Of course I admit it!" I say. Now Josh tells *me* to stay calm. "I have nothing to hide. I was in Baghdad writing articles about my own government's actions. I exposed U.S. military corruption and their support of militias." It's the same old mantra.

Our lawyer jumps in. "Your Honor," he says in a conciliatory tone, "like you and I, my clients are critical of their government's policies toward the Middle East. The whole world knows they are not spies and they should be released on compassionate—"

"What are you basing that on?" the judge snaps.

"If you watch TV, you can see people speaking about them all the time. Even famous Muslims like Muhammad Ali have said these two are not spies. Noam Chomsky has said the same thing."

"Do *you* believe what the Western media says?" the judge counters. "And what does it matter what these people say? We don't know who they are." Apparently he doesn't watch Iran's state TV, where a Muhammad Ali documentary and video clips from Noam Chomsky air frequently. "We judge people by their actions," the judge says, looking pointedly at Shafii, "not by words."

"Are you saying that because I am defending my clients, I am going to be marked as a spy too?" Shafii says. "This court isn't proceeding fairly. You think of me so badly that when I came into the courtroom you had a special security guard search my underwear and even the strap of my glasses." I can't believe how bold this man is. Lawyers get locked up all the time here. Is he *asking* to go to prison?

"We will tell you later why we searched you," the judge says menacingly.

But Shafii doesn't back down. "This is not supposed to be a public hearing," he says, looking toward the gallery. "So who are these fifteen people sitting here watching?"

"Those people are from the Ministry of Intelligence," the judge says.

"If Intelligence can come, why can't the Swiss ambassador?"

The judge orders him to sit down.

"Just one more thing," Shafii says, and pulls out a piece of paper. "This is a photocopy of the passport of Shon Meckfessel, the person that was with the three in Kurdistan. The visa in this passport explicitly states that it is only valid for Kurdistan, not other parts of Iraq. They could not have traveled to Baghdad with this visa. Shane's and Josh's visas are the same as this. You have not let me see their passports — you have them — so I have to present this."

The judge takes the paper and says, "Sit down. You have wasted enough of the court's time. We don't need to hear any more from you."

I am given a piece of paper to write about my previous trip to Baghdad. I answer it briefly, then attempt to describe how it would have been impossible for us to travel from the Turkish border to Baghdad and then back to Kurdistan and to the Iranian border in two and a half days. I look up and see the judge leaning back in his chair and smirking. When I finish, he turns to Shafii. "You say *you* are anti-American, but *you* don't have personal sanctions against you. America put sanctions on *me*. I am not allowed to travel and they are taking my assets." He wears his sanctions like a badge of honor. (I later learn that the sanctions actually came from the European Union, not the United States.)

The judge tells us to write our closing statements. As the translator reads mine, I look directly at the judge. He stares ahead blankly.

"'I am not a spy,'" the translator reads. "'In fact, I wish this hostility between my country and Iran would end. I am opposed to sanctions on Iran because they harm regular people. But I have no control over what my government does, just as I have no power to stop the wars in Iraq and Afghanistan. Even so, I apologize for these actions. I am sorry for these hostilities, including the sanctions against your country, including the sanctions against you, Your Honor. I wish people like you and I weren't caught in the middle.'" My statement has no visible effect on the judge.

He tells us he will issue his verdict within a week. Our guards usher us out. As we go, Shafii approaches. Josh and I look at each other, communicating with our eyes that it is too risky to pass the note. Shafii extends his hand to each of us, smiles, and utters four familiar words from a Dylan song: "'the ship comes in.'"

That's the song that we used to sing when we imagined freedom. Shafii knows. He thinks we are going home soon.

102. SHANE

Twenty days have passed since our last hearing. We are back at the courthouse, but this time we are taken into an office, not the courtroom. I sit down and Josh goes off to the bathroom with a guard to move our secret note from his underwear to his pocket. Shortly after he leaves, I am brought into another office with shelves full of legal books where the judge is sitting behind a court bench that is awkwardly tall for this small room. He doesn't look at me as I enter.

There are four men in the room besides the judge and the two guards with me. Our lawyer isn't here.

Josh comes back and sits next to me. "No lawyer," I tell him. "They say he's not coming. They say he will be informed of our sentence within twenty-four hours."

Suddenly, the judge starts speaking and our interpreter, the same man from our two hearings, starts writing. The judge is citing the legal code. We stare at the interpreter's notebook as he writes a bunch of Farsi.

The judge has his cell phone to his ear. He goes back and forth between issuing bits and pieces of our sentence in a loud, official voice and muttering asides into his phone. It is as if his issuing of our sentence is just some task he needs to finish up. No one in the room seems to be taking any of this too seriously. One of the men in the chairs along the bookcases is picking his nails. A turbaned man sitting up front is rolling his prayer beads through his fingers and looking out the window distractedly. Another man keeps yawning and staring at the ceiling. Their lack of concern is actually reassuring, as though everyone is going out of his way to acknowledge that this is merely a formality we need to go through, that it isn't serious.

The judge announces more words. The interpreter jots, then reads aloud: "You are each to serve a three-year term in prison for illegal entry. You each are to serve a five-year term in prison for espionage. That is eight years for each of you. Not both. Each. Eight years each."

"I would like to say something to the judge," I say to the interpreter. He tries to get the judge's attention, but he is still on his cell phone. The interpreter keeps raising his hand shyly, then taking it down. Eventually, the judge gets off the phone.

"Your Honor," I say, "I have been diagnosed with a serious health problem. I have an ulcer that is bleeding and it has made me anemic. I have not been able to receive the care I need in prison. This is serious and I need outside medical attention —"

"We'll make sure you get what you need," he says abruptly and with a distinctly unconcerned tone.

"No, you don't understand. The doctor has specified certain requirements for me, like a regular diet of lamb, which the prison has not been able to meet. I would like to request that you release Josh and me on humanitarian grounds like you did with Sarah so that I can get the medical attention I need."

"You want me to release *both* of you for *your* health problem?" he says, suddenly angry. "Listen to me right now. What happened with Sarah will *never* happen with you. Sarah Shourd broke the law by not returning for court. And I *guarantee* you that one day, she will sit right where you are and receive her sentence." I'm not really sure why he is saying this — he can't actually believe it. Is he putting on a show for his colleagues? Is he frustrated with his powerlessness in our case, knowing that ultimately, our fate is in the hands of politicians? "Twenty days from today, your lawyer has the right to appeal," he says. "This is our judicial process. Beyond that, you have nothing. Now go."

As we walk out, he shouts for us to stop. "What does he have there?" he says to the guards, pointing at a piece of fabric sticking out of Josh's shirt pocket.

Josh grabs the fabric between his thumb and forefinger and dangles it in front of the judge. "This is my blindfold," he says, half smirking. "We have to wear these in your prison."

"Okay, get out," the judge snaps.

When we walk out into the hall, there is a bounce in our steps. After two years, we finally got our sentence. Freedom must be around the corner. I feel almost giddy, but I try to hide it from the guards.

"Eight years," I say soberly to them as we go down the elevator. "What do you think?"

One of them brushes his hand through the air. "Don't worry about it," he says. "After twenty days, you will appeal. Then they'll let you go. This is normal. Don't worry about it."

103. SARAH

The verdict is all over the news. "The two Americans, Josh Fattal and Shane Bauer, have been found guilty on all charges and sentenced to eight years — three years for illegal entry and five for espionage."

Immediately, the families and I jump on a conference call with our lawyer. In terms of Iranian law, Masoud tells us, the verdict makes no sense. Punishment for illegal entry should be a fine with no prison time at all, unless it's a second offense, which holds a sentence of two years, not three. Five years is absurdly low for espionage, which should be punishable by death. Iranian American Roxana Saberi got eight years for espionage; then she was released after an appeal. Five years, we all agree, is the minimum the courts could get away with without openly admitting Shane and Josh aren't spies. Our lawyer tells us he will immediately file an appeal. We have to act outraged, he advises, but there is no denying this is a very good sign.

"This is awesome," Josh's dad, Jacob, says in his typically candid style. "It's finished."

Most of our thousands of supporters, lacking insights like the ones

Masoud offers, can't help but take everything they read in the news at face value. I turn on my computer and watch as hundreds upon hundreds of e-mails and Facebook messages flood in from around the world. "I'm so angry," one woman types on Facebook. "These people are sick. How can they hold Shane and Josh for six more years?" I can't exactly tell people they are wrong—the sentence is an outrage—but they don't know what we know.

A few weeks ago, Salem told me he had a *fatwa* from the Supreme Leader, a written document ordering their release. When Ayatollah Khamenei makes a decision on a legal matter, it's considered binding under Islamic law. For this reason, he doesn't often put things in writing. This time, the Supreme Leader sent a letter to Sultan Qaboos bin Said, which Salem will have in hand when he flies to Tehran. Now that the legal proceedings have taken their course and the hard-liners have given the conviction they wanted, there's nothing to stop a compassionate release. I hold this knowledge close to my chest and smile at the sky. Josh and Shane will soon be free.

104. JOSH

Judge Salavati handed us our sentence a few days ago. We're now waiting the requisite twenty days to appeal. Whatever whirlwind is going on out there, life in cell 111 is unchanged. We still lie on our beds and read all day, and we're still preoccupied with the letters that Dumb Guy doesn't bring. So once again, we decide to hunger-strike.

During our previous hunger strike, I ate, and I justified eating by remembering how the guards and interrogators often lied to me. But trying to deceive them affected how I related to them and, more importantly, how I related to myself. I know Gandhi would've judged me as a coward for secretly eating during my hunger strike. He believed Truth is Love is God. Since the fake hunger strike last April, a still small voice nagged at me.

This time Shane and I solemnly gather all our fruit, vegetables, and packaged food into several grocery bags. We place them in the hallway to prove to the guards that we won't eat. This time, I feel guilt-free and righteous—almost excited to be hungry.

The next day, Shane and I sit near each other on the floor. We take a "lunch break" even though we don't actually eat.

Suddenly, dress shoes clomp down the hallway and Dumb Guy arrives at the door. He tells us that there are no new letters for us, that our family hasn't sent anything recently, and that we should end our hunger strike. We refuse to eat without new letters, reasoning that he's probably lying again. Normally though, he takes at least three or four days to show up. This time he has responded in less than twenty-four hours. His appearance is a good sign.

A half-hour later, an administrator arrives at the door. Last year, this guy supported and sympathized with me after AK pushed me down the stairs. He now insists that Shane and I end our hunger strike. He promises that there are no new letters from the Foreign Ministry and he urges us to trust him. In less than one day, our interrogator made an appearance, and now an administrator has too. Something has changed. Their attention makes me more hopeful about our upcoming appeal.

Something has changed inside me as well. I let go of our ideology of resistance. Shane and I carry the plastic bags full of food in from the hallway to end our hunger strike. Even though we didn't win any letters, the process feels better to me. I no longer fixate on the narrow goal of defeating Dumb Guy. I am just trying to stay true to myself.

A few days later, Dumb Guy shows up unexpectedly with packages of letters. Our letters prove him right. Our families mention that they haven't written as regularly as they used to. In part this is because they're so busy with their efforts to free us.

105. JOSH

It's September 13, one day before the anniversary of Sarah's release. A guard I despise comes to the door. I once heard him tell Dumb Guy that we eat during our hunger strikes, but then he acts chummy with us when the interrogators aren't around. He glances suspiciously around the cell. *"Che tori?"* he asks. Why is he asking how I'm doing? I'm fine. *"Khordi?"* he asks. Why is he asking if I ate? Of course I ate. Since when has he decided to be so friendly? *"Television?"* He points to the TV. No. We aren't watching TV. We don't turn it on until the English news later at night. I wish he'd leave us alone.

He again points to the television and yaps away in Farsi. I'm losing patience with him, but then I hear the words *azadi, freedom,* and *shoma, you.* Then he points to the TV and says, *"Ahmadinejad goft,"* *Ahmadinejad said.*

When he sees that we understand him, the guard dons a rare grin. I flip on the TV, and Shane grills him. When exactly will we be freed? Which channel is it on? This is not something to joke about! He promises he's not joking and points to the TV. *"Emshap,"* tonight, he says, then leaves.

Shane and I watch the English news ticker on IRINN: "Iranians Make Scientific Advances Despite Sanctions. World Powers Face Collapse Due to Their Hypocrisy. IAEA Says They Respect Nation's Right to a Peaceful Nuclear Energy Program." It's the same old blah-blah-blah. Nothing about us. I change the channel and flip through the Farsi news. Nothing about us. The fifteen minutes of English news is on early tonight at 7:30 p.m. Nothing.

After an hour, I unglue myself, clean up dinner, and resume reading about the Napoleonic Wars. Shane attentively flips between four potential stations, making sure not to miss any short news segments. We ask other guards to ask that guard to return, but he never comes.

"Could he have been messing with us?" Shane asks me.

"No way. Even he wouldn't make that up," I say with as much confidence as I can muster.

Every twenty or thirty minutes, Shane checks again. Clinging to my belief that the guard wasn't lying, I continue reading while Shane stares at the TV. Every hour or two, I scan the channels with Shane. There is still nothing about us.

Though it's past midnight, we're still watching the English news. Still nothing.

The news ends and Shane goes into the bathroom. I watch one more cycle of the ticker before trying to sleep. It is a rare moment when I'm watching the headlines and Shane is not.

"Hey, Shane," I say matter-of-factly, "Ahmadinejad says that we live in 'hotel-like conditions.'"

Shane lets out a small chuckle from behind the bathroom door. "Hey, Shane." I read him the next headline in the same even voice: "'Ahmadinejad Says U.S. Nationals to Be Released in Days.'"

He reappears. "You weren't kidding!"

We watch the headlines flow past, and when Shane finally reads the ticker, we give each other a huge hug, laughing in each other's arms. Tears come to my eyes.

106. SHANE

A week later, Ehsan comes to the door. *Is it time? Are we going?* I've always imagined Ehsan escorting us out of the prison for the last time. I've pictured Josh and me standing at the door of Section 209, smiling at Ehsan while he beams his huge smile back at us and shakes our hands. I would thank him for everything he has done for us and he would shake our hands and congratulate us, heartily, for our freedom. Then, after we are free, he might even contact us. Maybe we'd become friends.

"Can you step outside, please?" Ehsan says to me.

My heart jumps. "Just me?" I say, and look at Josh. Is this where I retreat deep into the cell and cling to the bedpost? Where they try to pull me out and leave Josh in here alone? Is this where they ruin our lives even further?

"Yes," he says. "Just you." I think he sees concern in my eyes. "Don't worry. You will come right back."

I exhale. I believe him. He is the one and only person who works here that I trust. I slip on my blindfold and step out. I see another pair of feet from under the blindfold. I tilt my head back to look. It's AK. I pull my blindfold up and look at Ehsan with a what-the-hell-is-going-on look.

"My friend would like to say something to you," Ehsan says. "He would like to apologize for what happened."

Apologize? For pummeling me a year ago?

AK looks at me sheepishly with his hands in his pockets, his head slightly hung. "I am sorry," he says in English, extending his hand. I take it. "Please excuse me."

I can't decide whether someone put him up to this or he decided to clear his conscience. I do know that it means we are going to be freed. The three of us stand there, staring at one another. Why isn't Ehsan saying anything? Why hasn't he told us we are getting out soon? Why was it a guard we dislike that slipped us the news? Ehsan has been so tightlipped. Actually, now that I think of it, he has never given us any information from the outside. And we've asked him for it dozens of times. I feel a little hurt at the thought that, when it comes down to it, Ehsan is more concerned with his position as a guard at a political prison than he is about its inmates. He is still one of them.

AK shakes my hand again. I can't help but smile a little. He smiles a little too. I am starting to feel betrayed by Ehsan, but in this moment, I forgive AK.

107. SARAH

I'm deep asleep when my cell phone rings.

"We are almost there, Sarah," Salem says. "It is time to make your move." Since the verdict was announced and the appeal filed, I've done nothing but wait for this call.

"Salem," I say, "I owe you my life and all my happiness forever."

"Your happiness is mine," he replies warmly.

Salem tells me the families and I should all get on the next flight to Oman. I ask him when he'll be leaving for Tehran. He laughs at my impatience, then tells me he's already en route and will be there in an hour.

"Now get your butt over here!" he says, laughing.

I hang up the phone and start laughing myself. It's still the middle of the night. Without thinking, I run into the bathroom. "You're going to Oman!" I mouth at my reflection in the mirror. I look into my own shocked, almost frightened eyes and begin to laugh again. *I'm going to Oman!* I shout.

I dial Cindy's number as I start throwing clothes into my huge, beat-up suitcase, with which I've crisscrossed the country dozens of times over the last year.

"Sarah?" she says, answering the phone. Even though it's 4 a.m. in Minnesota, Cindy sounds alert. I hear in her voice a readiness for anything that I might say.

"Salem asked us to come to Oman, Cindy. He's on his way to get them."

Ten hours later I'm in the Omani airport, waiting as Shane's and Josh's family members trickle in from different flights. It's been exactly one year and one day since my release into this sparkling world of turquoise and white, and now I'm back. Al is all smiles and Laura is chatty as usual. Cindy looks weary but strong, Alex is giddy, and Nicole and Shannon are nervous and tired.

We all pile into a van and are driven to the ambassador's residence, where Richard Schmierer and his wife greet us at the same mansion I stayed at last year. Everyone disperses into their rooms, but I know I'm not going to sleep, so I wander downstairs to the kitchen. The same, sweet maid from last year greets me with a warm hug and hands me a plate of apple pie. I wander outside the embassy walls and find the beefy former marine at his post. He offers to accompany me on a jog, setting his pace to my own and hanging back at a respectful distance.

My feet rhythmically pound the warm sand. Nothing here seems to have changed, but I feel deeply changed. In my mind's eye, I can see Salem walking through the gates of Evin Prison in his long robes and headdress, carrying a suitcase full of cash with which he will pay the million-dollar bail. This last year has been one of the most difficult, yet empowering, years of my life, and I feel proud of what I've been able to accomplish. No experience, I realize, could have shown me better who I am.

We spend the next week tiptoeing around one another. With eight family members in one house, we're lucky the residence is spacious and the ocean is right out the door. Shane's dad, Al, and I often spend the morning sitting and talking under a palm tree on the beach until the sun gets too hot and we have to go inside. One afternoon, all eight of us take a special tour of the Sultan Qaboos Grand Mosque, by far the most gorgeous building I've ever seen. It's built of Indian sandstone, with a huge golden dome in the center and four flanking minarets surrounded by fountains and lush, green gardens. We eat out every chance we get, go shopping, watch the news, and wait for Salem's coded texts.

"I almost have all the ingredients and *inshallah* tomorrow I will bake an apple pie for you. Keep a pot of hot coffee ready."

Everything is playing out almost exactly like it did for my release last year, except this time there are even more complications. Every time the president makes a move, the judiciary seems to find a way to block it. As soon as Ahmadinejad publicly announces his intention to release Shane and Josh, the judiciary predictably counters by demanding a million dollars in bail. Then, they throw in a new twist. When Salem shows up with the cash, he is told that the bail can't be accepted until a mysteriously missing judge gets back from vacation. A day later, the judge is still missing.

Then, on September 21, 2011, at 8:42 p.m., the text we've been waiting for finally arrives:

Salem: We are coming home.
Ambassador: Confusion abounds. Released?
Salem: No confusion. *Inshallah* we will be back tonight.
　　See you at 10:30 p.m.

Shortly after the exchange, we read in the news that the missing judge has surfaced, the bail money has been accepted, and the release has been signed. Then, we get a call from Josh and Shane, about to board Salem's plane at Tehran Airport. I turn on the news to look for footage of their departure, knowing that our friends, family, and supporters around the world are doing the same. At that moment, all the news channels are covering President Obama's United Nations address. Below the president's face on the screen a ticker reads:

"Breaking News: After two years in prison, Americans Shane Bauer and Josh Fattal are released from Iranian prison on grounds of compassion after being sentenced to eight years for illegal entry and espionage."

Suddenly, the cameras cut away from Obama, standing at the podium in front of the United Nations Assembly, and zero in on a balcony on the other side of the room. President Ahmadinejad himself is standing there, flanked by his entourage, smiling and waving to the audience.

"I'm the good guy here," President Ahmadinejad seems to be saying with his innocent smile. "I let Josh, Shane, and Sarah go. Why does everyone demonize me?" The Iranian president must have planned to interrupt Obama's speech just as Shane and Josh are being flown out of Iranian airspace, but why?

No Iranian or American president has attended his counterpart's speech at the UN in recent history. As Shane and Josh mark the end of twenty-six months in prison, President Ahmadinejad has boldly placed himself in the same room as President Obama, the closest these two men will ever be.

The moment seems largely lost on the media. The camera lingers on the Iranian president for only a few seconds before returning to Obama, who hasn't even deemed it necessary to pause in his speech. I hear footsteps on the stairs. An embassy official rushes in to tell me the van is waiting outside.

108. SHANE

As Josh and I walk to our cell from *hava khori,* someone redirects us into the doctor's office. We take off our blindfolds. The doctor smiles. "Please sit," he says.

He takes our blood pressure.

"What is this?" Josh asks.

"It's your annual checkup," the doctor says in his lilting English. "To make sure you are healthy."

He puts us on the scale. I am twenty pounds lighter than I was when we got here.

Josh keeps eyeing me and I keep avoiding his gaze. I'm not ready to acknowledge what I know he is trying to suggest. We'll know it when the time comes. Let's not jump the gun.

When we leave the doctor's office, we aren't taken back to our cell. Two guards whisk us down the stairs and out the front door of Section 209. Dumb Guy is there, looking frazzled. One of the high-level administrators steps out of the building behind us, grabs the books we happened to be reading at *hava khori,* and disappears back inside the building.

Dumb Guy rushes us into the storage room across from Section 209 — the same place where we change into our street clothes whenever they take us outside. As soon as we enter, I see all of our bags spread across the floor. Finally, it starts to sink in.

"Guys, take your things," Dumb Guy says. "Make sure everything is there and sign this paper. Quickly! The Swiss have been waiting for a very long time."

Josh and I dig through our bags. I search for my camera and money. It's all there. In fact, there's much more here than we came with. There are bags full of the dates — now rotting — that Salem brought us over the months, brand-new clothes we never received, razors, shaving cream, shampoo, and a dozen little battery-powered fans, surely sent by Josh's mom. Suddenly, I fear that we're going to lose everything. None of our books are here. We tried to donate them to the prison library for other prisoners to read, but they wouldn't allow it. They should be here. There should be hundreds. Not only have these books been our lifeline, but our secret journals are tucked in their spines. We can't let them go.

"What about everything in our cell?" I ask. "Where are the books? You have to give us the books."

"We will mail them to you."

"No, we need them now," I say.

"Books?!" he snaps. "Are the books important right now?"

I have hated this man, but I don't think hate is what I feel anymore. I haven't quite felt that in a while, as much as I have wanted to. He has been the face of the regime holding us captive for so long. He has messed with us and made us suffer. He has neglected us. But he has also been the one to bring us things, to ask if we have any problems, to make sure we aren't being abused by guards, to give us birthday cakes, to take us out to a park. Now, it feels like we are a bickering couple about to sign divorce papers on terms I expected to be amicable, but aren't.

He looks me in the eye. "You have no idea how much I have done for you, Shane. You are being released right now because of me." He seems to really believe that.

We pile all of our things into the trunk of a car. Guards shake our hands and give us hearty goodbyes. When we get in the back seat, Josh grips my knee, smiling. Today is September 21, our 781st day in Iran.

109. JOSH

Dumb Guy eventually takes us to another building within the prison compound. I'm wearing a brand-new collared shirt and the fancy watch that Salem gave me on his first visit more than a year ago. I notice that the second hand isn't ticking.

A gaggle of people crowds the doorway of this building. I catch the familiar scent of sandalwood. I know that smell. I rise to my tiptoes to look above the crowd. I see Salem al-Ismaily up ahead in a dark turban. I weave through the stagnant mass and make my way to him. He's dressed in a white, full-length traditional Omani robe.

He grabs my arm and pulls me to his warm body. He speaks the words I've been longing to hear—"Let's go home."

The next thing I know, I'm sitting in the adjacent room at a table with candies, a mini Iranian flag, and a microphone in my face. Shane sits by my side, and a video camera with the IRIB logo points at us. A dozen people are standing around and the interviewer starts the questions:

"How were you treated?"

"Not well," Shane says. "We have been isolated for twenty-six months. We haven't been allowed to meet with our lawyer. And we are innocent. We should have never been held here."

"Were you tortured?"

"We were held without any rights and extremely isolated," I say.

"Are you grateful to President Ahmadinejad or Ayatollah Khamenei?"

"Nope," Shane says.

"Don't you appreciate the compassion of the Iranian government for releasing you before your eight-year sentence is up?"

"No. This isn't compassion! This is a political prison and you held us for your political ends. You know we're innocent," I say, getting worked up.

The crowd is getting worked up too. One guy storms out. Others start whispering to one another. The interviewer's tone grows frustrated.

"How was the food? Don't you like Iranian food?" he asks, grasping for an innocuous topic.

I stammer a moment, thinking of the delicious lamb and dates. But I don't want to flatter them. "It was so-so."

"Did you learn any Farsi?"

"Yes!" I say with a smile for the camera. "I learned a little bit of Farsi." The interviewer takes a hopeful step toward me. I look around the room and say, "The guards taught me: *Sari! Vaisa! Boro! Cheshband paa'iin!* and *Fardo!" Faster! Stop! Go! Bring your blindfold down!* and *Tomorrow!"*

The interviewer sets down the microphone and turns off the camera. As I exit the room, I see Dumb Guy in the corner shaking his head, looking embarrassed.

110. SHANE

We get into a black car with tinted windows. The gates open and the gates close. We float past all the people waiting outside, in the dark, hoping to see, or find, their loved ones. As we roll down the highway, Josh and I talk and laugh. The driver, his brow furrowed, stares at us in the rearview mirror. We look back and smile.

An hour later, we board a huge passenger plane where we and our four Omani escorts are the only passengers. After takeoff, we are feted with gourmet Arabic food. Salem tells me he has a priest waiting to marry Sarah and me. He boasts to us about his villa on the beach, which is waiting for us. Then, as though he is telling me of a dying friend, he says that Syria is at war with itself. Our old home, the place where I spent one of the happiest years of my life, is now being ripped to shreds by bombings and sniping.

I dig a razor out of one of our bags and go into the lavatory to shave. I don't want to come out of prison looking haggard. The act of running the razor over my skin is gratifying. My face hasn't been this smooth and clean in more than two years. I spray on the airplane cologne. I try to picture Sarah and our families, but the lens is cloudy. I wonder what the Iranians will do with our secret journals once they find them. I wonder how they will react when they discover all the brewing wine in our bathroom. I wonder what will happen when we land.

The airplane touches down in Muscat. A group emerges from the airport and walks toward the plane. Josh and I strain to pick out faces we know. Everyone is so small. I think I see them, walking from the little airport in front of a row of cameramen and soldiers. They're waving. They must see our faces framed in the little windows. My heart leaps. This is for real.

111. SARAH

A small crowd has assembled on the tarmac at Muscat Airport. Salem's private plane has just landed and parked twenty feet in front of us. I can barely make out a shadowy figure behind one of its small windows. I don't dare open my mouth or move my eyes from that window. Suddenly, the figure waves and a cry escapes from my mouth. It must be one of them. I know it's them.

"Where's Sarah?" I hear someone ask, and I'm pushed to the front of the crowd. There are dozens of people — family, reporters, and Omani government representatives standing all around me. I look at them, but I don't see them. Someone wheels the stairs up to the door of the plane. Seconds pass and nothing happens. Then the door opens, forming an arc of bright light.

Two bodies appear backlit at the top of the stairs.

112. SHANE

As we arrive at the door, I don't know if I am entering a dream or leaving one.

The crowd at the bottom of the stairs sways and distorts in front of me. As I descend, running, I search for Sarah. At the bottom of the stairs, we collide. She is in my arms and I am kissing her. "You are my hero!" I whisper into her ear, turning her in circles. I see Mom smiling, standing back quietly. I go to her, give her a flower I picked from a bouquet on the plane, and hug her tightly, laughing from deep inside. Everyone is gushing emotion like I have never seen. Dad is weeping so hard, it is convulsing his large body. Shannon is smiling as if possessed. Nicole's sobs contort her face with relief so enormous that she couldn't have known she was carrying it. I hear Laura screaming, but I can't see her. I go from one person to the next, embracing each. Sarah is always in front of me, looking at me with eyes big and beautiful. She is not like she was. She is alive again. I hand her a flower, then scoop my sisters into my arms.

113. JOSH

Alex squeezes me tight. My momentum pushes him backward into Mom and Dad. All of their arms surround me at once. I grab my father and pull him closer — tears streaming down his face. My mother's lips are on my cheek, then my forehead, and I hear her screaming, "You made it! You made it!"

"Mom, *we* made it," I say, pressing our bodies together. "*We* made it."

I think of the 781 letters she sent me; of Alex crying at the vigils; and of Dad blaming himself for my imprisonment. I clutch each one and don't let go.

We continue hugging and kissing until a cameraman interrupts and asks for a photo. We pose together. Sarah watches my family, her cheeks wet with tears. When I see her smiling face, I rush out of the photograph. She lets out a yelp when I hug her. Swaying back and forth with her on the warm tarmac, I close my eyes and rest my head on her shoulder. I lean back to look her in the eyes. I stay in Sarah's arms until security guards usher us indoors.

In the airport, everyone stands together in a circle and the ambassador tells us that the media wants a statement from Shane and me. A discussion ensues; everyone has an opinion: Should they speak? What should they say? Should a family member make a statement for them? I feel overwhelmed.

"They're free now," Sarah says. "Let them speak for themselves."

"Anyone have a pen?" I call out.

One of the U.S. embassy staffers instantly hands me pen and paper. I twirl the pen in my hand, remembering the months of nagging the interrogators for one. I imagine us hiding in the corner of the room every time we'd write with a contraband pen.

Shane and I huddle to write a brief statement to the press. Salem asks us to thank the Supreme Leader, but we don't want to do it. Together, we draft four sentences, but Shane and I both want to say the same two lines. We're familiar with this kind of conflict. Shane wins the coin toss.

We sit down to practice our speech. Our families are looking at us and talking, trying to figure out how damaged we are. My brother videotapes Shane and me rehearsing.

"Josh, you keep forgetting to say 'His Majesty.' It's 'Our thanks to *His Majesty* Sultan Qaboos of Oman.' Okay?"

We get in front of the cameras and I say, "We are so happy we are free and so relieved we are free. Our deepest gratitude goes toward His Majesty Sultan Qaboos of Oman for obtaining our release. We are sincerely grateful to the government of Oman for hosting us and our families." Shane adds, "Two years in prison is too long, and we sincerely hope for the freedom of other political prisoners and other unjustly imprisoned people in America and Iran."

A few buff, straight-faced security guards usher us into cars and to the American ambassador's residence.

As soon as I enter the residence, a young woman pulls me aside. She has a sweet face and long, black, shiny hair. She asks if I'd follow her into a private room.

In the room, Shane is already seated and waiting. She tells us that she works for the U.S. State Department, and asks us to sign some papers. She offers to get us anything we need. Then she asks if we will speak to an FBI agent. Shane and I hesitate.

"Maybe you know something that will be useful," she suggests.

"We don't know anything. We were highly isolated and couldn't speak to people," Shane says.

"We think that Iran has been holding another American prisoner incommunicado since 2007."

"We didn't have contact with any Americans," I say. "And nobody mentioned another American to us."

Shane and I consult privately about what to do. Neither of us wants to meet with the FBI. We've already told them we don't have any information. What else do they possibly want? Before prison, we would never have collaborated with U.S. intelligence agencies. We won't now either. I don't want to play into the mutual hostility that created this mess or take sides between the U.S. and Iranian governments.

114. SHANE

I get up and walk around the house. Everything seems so familiar, but somehow disjointed in my mind. There are bedrooms with ruffled suitcases, open laptops, a treadmill, a big TV, books on random tables. Everything carries some profundity that I can't place. I go into the kitchen and open the cupboards. I pour salsa into a bowl and scoop it up with chips. Then I have a bowl of cereal followed by cookies and ice cream. My mom and sisters smile as they watch me eat. I can see myself being watched. I can see that everyone is more sophisticated than I am. They can talk to me and observe me and one another all at the same time. For the first time in my life, I feel like I can't do that. I can see that people are communicating with one another in many subtle, nonverbal ways, but I don't understand them. Josh I can understand. I can watch him and see his happiness, his confusion, and his uncertainty. I can read the lines of Josh's eyes, the turns of his mouth. Everyone else's body language is impenetrable to me.

Sarah guides me through the hours, motioning for me to talk to my dad or to go to my sister when she is crying. It's like she knows something about what is happening to me that I don't. It's like she can think for me.

A group of us goes up to the roof. It's dark outside. From the edge, I can see waves lapping onto a shoreline illuminated by the glow of buildings and streetlamps. A brightly lit ship floats far out in the blackness. I put my arm around Sarah. Her head rests on my chest. The air smells of ocean and apple-scented smoke. Nearby, a man is sitting outside a shop and smoking a water pipe. The waiter comes and goes, and they engage in the quiet conversation of friends. A warm sensation of familiarity washes over me. I realize that I have forgotten the way that time passes out here, the evening walks down the street, the after-dinner smokes and thoughts. Then, I am seized by a desire.

"Let's go out there," I say to Sarah and Josh in a tone that suggests I am proposing something daring. I think of the woman Josh and I spoke to. She suggested we shouldn't go outside.

"Are they holding us prisoner?" I say to Sarah and Josh. "We are free. Finally, we can go outside. I want to go out and walk." I am trying to convince them that we can go outside, that we should insist on it. I don't realize that this is the only way I know to get what I want. I have trained myself to be either acquiescent or defiant, with little in between. I don't understand that I am actually free, and if I want to, I can just walk outside.

Sarah smiles. "Let's do it," she says.

115. JOSH

The water is alive. It actually sparkles. With water up to my thighs, I make waves of glittering light with every slow, deliberate step I take. I skim the surface with my hands, creating ephemeral designs of light in my fingers' wake. The bioluminescent water twinkles like the stars above.

I lean forward and plunge the rest of my body into the warm sea. I feel buoyant — able to swim with ease, held by the ocean. In prison, I read a book called *Big History*, which said that there is the same percentage of salt in seawater as there is in tears. I think of the Passover ritual of tasting salt water to remember the tears of oppression. I imagine this ocean to be trillions of tears — mainly sorrowful tears from family and friends, from Shane and Sarah, and from my eyes. As I swim farther from the shore, I imagine a splash of joyful tears giving blessings to these dark waters.

Turning around, I face the sky, and I float between the starry night and twinkling ocean. There is splashing nearby. I hear Nicole, Shannon, and Alex giggling. Shane and Sarah swim together toward the horizon. My mother stands shin-deep in the water, holding towels in her arms so we won't be cold when we get out.

I swim back toward the shore, and Sarah swims beside me. The water is only a couple feet deep, but I'm not ready for dry land. I stay horizontal, holding the sand with my hands as I let my body ebb and flow with the tide. I turn to Sarah. I've been reading her letters for the past year, but she has barely heard a word from me.

"I've so much to tell you, Sarah. You've meant so much to me." She listens intently. "I thought of you so much. Your letters were so special to me. I feel like I understood you even more after you left. I came to appreciate how you held us three together, how strong you were. I've been wanting to tell you that for a year."

"You mean so much to me too!" Sarah says. "This is just the beginning of your journey, Josh. There is so much that I want to share with you! When we were in prison, you were so there for me. Campaigning this past year felt like my way to make it up to you. This is just the beginning, Josh — remember that. It takes a while to adjust to freedom. Take it slow."

Sarah asks about the guards and the interrogators. I'd like to stay in these shallow waters and catch up with her for hours. But the rest of our group beckons us to walk back to the residence.

I float along with the tide for another moment. *Big History* also states that the percentage of salt in seawater is the same as the salt in a mother's womb. I step out of the coruscating Persian Gulf and feel myself reborn.

116. SHANE

Hours pass and the sky begins to light up. The three of us go back up to the roof, where Sarah hands me a bottle of champagne. I shoot the cork off into the air and pour three glasses. We toast and look at one another, *into* one another. We don't speak at first. We sip and look out to the place where the sky and ocean meet. The cold ocean depths touch the air, which is the beginning of the sky, and the sky never ends. I take another sip and look in the other direction. The narrow corridor of houses is hemmed in between the shoreline and a range of short, rounded mountains. The three of us sit down and hold one another's hands, resuming the position we used to get in at *hava khori*, our way of giving one another strength and reassuring one another that everything was okay.

Sarah breaks the silence. "I have a song I wrote for you guys," she says. "I wrote it when Salem told me in July that the Supreme Leader said he would release you." She sits up straight, closes her eyes, and takes a deep breath. Josh and I don't lie down and look at the sky like we used to when Sarah sang. We sit up with her, facing her.

The words roll out of her. "I'm not free until you're free. And we're not free until they're free." The sun edges over the mountain and tears run down my face. Sarah is crying too as she sings about tearing down prison walls and dancing and not being afraid. Josh's eyes are dreamy. I don't know whether what I am feeling is sadness, happiness, relief, love, or loss. It is like all emotions are piled on top of one another, like some part of me that has been tucked deeply away has been suddenly unlocked and its contents are now pouring out. Everything is making me cry: the growing warmth of the sun on my skin; the small, elegant clouds that texture the bright blue sky; the flock of blackbirds that fly between that sky and us; the sound of Sarah's voice and the love and sorrow it contains.

When she stops, the three of us fall toward one another. Josh squeezes my shoulder. Sarah runs her fingers through my hair. We hold one another, and we cry. And we laugh.

SARAH

"I feel I deserve to know," I type. "Where exactly is the border?"

It's been almost three and a half years since the day we were captured by Iranian soldiers. I've played and replayed the events of that day countless times in my head. Now, I'm sitting behind my desk at my office in Oakland, California, chatting with an anonymous man on Facebook who says he's one of the soldiers who drove us into Iran.

"What about the scarf?" he asks, eluding first my question. "Do you still have the adventure scarf?"

He's referring to the headscarf the soldiers stopped to buy me on the way to the police headquarters in Mariwan. I write back that no, I wasn't allowed to keep the headscarf. Still, that detail alone doesn't prove who he is, since I've mentioned it in the media many times.

"Remember WC?" he types. "I say WC and you smile."

I do remember WC. About an hour after we were detained, the three of us were sitting on the porch of the small military shack. Dozens of curious soldiers were clustered around us, trying in vain to communicate while their commander waited for the call from Tehran that would determine our fate. I asked if I could use the bathroom.

"Water closet?" one soldier responded. Language classes in Iran must teach British English instead of American, I thought.

"WC?" he said again. "Yes. Come with me."

Now, persuaded he's telling me the truth, I type, "Please why won't you answer me? Where is the border between Iran and Iraq? I have had a very bad time in your country; I deserve to know."

"I am very sorry," he types back. "I hope you are better now. What is your question? I don't understand."

"Is the line between Iran and Iraq (a) the path, (b) the spring, (c) the round building, (d) the ridge?"

"It is C," he says, confirming our long-held belief that we were lured over the Iran-Iraq border deliberately. The soldiers didn't want to risk crossing into Iraq to approach us, so they motioned to us from the top of the ridge and — from their vantage point at the building — watched as we unknowingly walked into Iran.

"Thank you," I type. "I appreciate your being honest with me."

It's taken me a long time to find closure. Five days after Josh and Shane were released, I woke up in a small bedroom decorated with antique Victorian dollhouse furniture. I could hear Shane and Josh talking in the kitchen downstairs — we'd come there with our families directly from the press conference Shane and Josh gave in New York. My first thought when I opened my eyes that morning was, *I need a new mission.* Now that Shane and Josh were free, and no one else's life or freedom depended on me, what was the purpose of *my* life? Without a new mission, I couldn't see the point in getting out of bed.

In the following weeks and months, I fluctuated between happiness, rage, and depression. One morning I woke curled up in a fetal position in the corner of a friend's guest bedroom, mortified by my partial recollection of horrible, spiteful things I said the night before after drinking two bottles of wine. Was I really that full of rage? Why did it only come out when I was drunk? I'd woken with a debilitating headache, and I was scheduled to meet a prospective book publisher on the opposite side of Manhattan in forty-five minutes.

Months later there was another incident that revealed a lot to me about just how much healing I was going to have to do. I was deboarding a plane when the power suddenly shut off. Paralyzed with fear, I was convinced we were all about to be gunned down. Before I could scream — I was going to tell everyone to "take cover" or something equally cinematic — the lights came back on, and the flight attendant spoke over the loudspeaker, apologizing for the inconvenience. Later, back in our apartment, I lay awake the entire night, too terrified to move, waiting for the cabdriver who drove us home from the airport to come back and kill us.

These delusions, drunken rages, and panic attacks scared me. I decided to limit my alcohol intake, lessen travel, start doing yoga — which helped with my breathing and anxiety — and spend as much time as possible reconnecting with people I love. Shane and I, sometimes frantically and other times gracefully, began the process of putting our lives back together. We filled our apartment with color and comfort, planted succulents, bought new bikes, stocked our kitchen, and began to plan our wedding and our future.

Years later, I'm still healing. Though my sleep patterns are back to normal and my panic attacks less and less frequent, no aspect of my life is or ever will be as it was. The most empowering and challenging part of free life has been choosing what parts of prison I want to integrate and what parts are best left behind. Solitary confinement opened up existential questions that were too uncomfortable to sit with, so in that situation I sought answers. Worship and the notion of a personified God helped me tremendously in prison. I needed a higher power I could speak to, a relationship through which I could orient myself to the world I'd lost. I even started using the pronoun "he" for a divine personality, which never made sense to me before and has no relevance now.

I still believe in an abstract higher power, a force and intelligence behind the interconnected nature of all things, but I'm no longer religious. Now, I meditate, I pray, I try to find new ways to respect and worship life every day. Basically, I try to be guided by a vision of a reality greater than myself. My spiritual path is by no means complete — I haven't figured everything out and frankly don't expect to — but I'm still on it. I've always been on it.

When I first got out of prison, some people accused me of being brainwashed. A few well-known bloggers said that my insistence on saying positive things about Islam and the Middle East could only be explained by what psychologists call Stockholm syndrome — feelings of trust or affection a hostage develops for her captors. This accusation was racist and ridiculous — I didn't have to be brainwashed to appreciate the beautiful, diverse cultures I'd experienced in the Middle East. I'd spent one of the best years of my life there before I was captured. Still, I've come to see that there were elements of my behavior in prison — particularly my obsession with the interrogator I revealingly called Father Guy and the period when I considered converting to Islam — that can be explained by this diagnosis. In an attempt to please my captors, to elicit their respect and mercy, I subconsciously tried to become more like them.

Once I was free, those symptoms passed very quickly — I remember not wanting to show my arms or legs in public at first; now I wear whatever I like — but it didn't take away my tremendous respect for the Muslim faith and love for the Middle East.

An important way to integrate my experience in Iranian prison has been connecting my own suffering with the suffering of others. Even after the recent election of a more moderate president, Hassan Rouhani, the Iranian government continues to do what it did to us, and much worse, in an effort to suppress the demands and aspirations of its own people. Being held hostage directly after the zenith of Iran's popular movement for democracy left me in awe of the determination of the Iranian people, including the women whose bold, beautiful faces I often saw pass by my cell at Evin Prison. Their courage made me glad that our story brought attention to this brutal regime, and hopefully even played some small part in speeding its downfall. Though the movement for democracy in Iran has been largely dormant in recent years, the extreme paranoia of the Iranian government and the brutal repression it continues to exact are an indication that it still feels threatened.

An attack on Iran by the United States or Israel would be a human rights disaster. It would embolden Iran's corrupt leaders to ramp up homegrown repression even more, forcing Iranian citizens to rally behind a government the majority of them despise. The Iranian people are centuries ahead of their leaders, politically and culturally, and their progress, which, like that of many other peoples that have recently sparked revolutions against brutal dictatorships in the Middle East, should happen on its own trajectory.

It's also important to understand that the inhumane conditions we've described in this book are far from unique to Iran. The year that I spent in solitary confinement was not living — it was a space between life and death, and it threatened to erase me. When I found out that the United States holds more people in solitary — up to eighty thousand on any given day — than any other country in the world, I was shocked and outraged. The UN has confirmed that such confinement for any period over two weeks can constitute torture. Yet we routinely hold people in isolation for months, years, even decades, often as punishment for nonviolent infractions. This draconian practice benefits no one — quite the opposite. Studies have shown that it increases violence in prisons as a whole and increases recidivism. Less than 5 percent of all prisoners are in solitary in the United States; yet as many as 50 percent of all suicides happen there. After being subjected to this cruel, inhumane treatment, many people are released directly into our streets—traumatized and completely ill-equipped to make better choices in the free world. Even if you wish to put aside a moral argument in favor of a purely pragmatic one, the widespread use of psychological torture hasn't proven effective for rehabilitation or made our streets any safer, so why are we allowing our prisons to use it?

Though it can be difficult for me to engage with the cruelty and inhumanity in my own backyard—sometimes triggering my own memories—it would be much more difficult not to. It's an important part of my healing, and it reminds me that this story isn't just about Shane, Josh, and me. It's about everyone whose lives we've touched, continue to touch, and those who've touched ours. A woman named Zahra Bahrami—who showed me love and kindness months before she was executed for a drug-related crime, one she most likely didn't commit and one that should never warrant the death penalty in the first place—changed my life. A man named Jafar Saidi—a former Black Panther who wrote the Supreme Leader of Iran a letter advocating for our release and miraculously got a response—is sitting in a prison somewhere in Pennsylvania that he will likely never leave. Our friends in Syria—now living as refugees scattered around the globe, many of their homes destroyed, people they love murdered by Assad's regime—showed me just how beautiful, open, and welcoming the Middle East can be. Lastly, I will always be grateful to the countless people around the world who believed in us, fought for us, and gave us our lives back.

Watching the sunrise will always remind me of the morning on the roof in Oman when my life began again. Sometimes even the simplest acts, like walking out the front door of my apartment into my bustling neighborhood, will remind me forcefully of what we've all been through over the last four years. In those moments of transition—from night to day, from restriction to openness—I recall what they took from me, what they could never take, and ultimately just how lucky I am.

JOSH

"Do you speak English?" I asked. The soldier nodded. I continued. "When will I get out of here?"

"Soon," he answered grumpily.

"You told me that hours ago."

"Be patient! The interrogator will call you when he's ready," the soldier responded.

I sat down and wondered what returning to an interrogation room would feel like. Would it feel familiar? Would it trigger a traumatic reaction?

I had yet to understand how I'd been affected by my time in prison. Within six months after my return to civilian life I came to believe myself fully acclimated. I stopped hoarding pens, collecting used Scotch tape, and marveling at the comfort of sheets. I stopped routinely losing my keys and locking myself out of my apartment. I stopped expecting the call to prayer at dusk; I broke my habit of greeting waiters with *"Salaam aleikum"* as I would prison guards when they brought me food. Yet, still, more than a year later, I often got nervous talking about my time in prison. Mental health professionals warned me that acclimation would take time. One professional advised me to avoid the Middle East, confined spaces, and armed guards. But here I was, back in the Middle East, talking to a soldier, and confined in a twenty-five-by-fifteen-foot holding pen.

I had just arrived at Ben Gurion Airport in Tel Aviv. The Israeli customs officer had scanned my passport and sent me to this nearby detention room. A soccer match flickered on a TV in the corner of the cell. Without phone coverage or Wi-Fi, I was unable to contact my father and my extended family waiting for me in baggage claim. Eight other travelers sat along the perimeter of the cell, staring in silence at the white walls. I felt like I knew a secret they weren't privy to: we weren't in solitary confinement — at least we had one another. I made a point to talk to each of them.

Three hours later, the grumpy soldier escorted me to the interrogation room. On the short walk there, I assured myself I'd be safe. After all, I had proper documentation to enter their country; the

U.S. and Israeli governments were friends; and as added security, the news media expected me to arrive.

"Why am I being detained?" I asked.

The interrogator swiveled in his office chair and typed loudly on his computer keys. He continued to ignore me as he flipped the pages of my passport. Eventually he turned back toward me and asked in accented English, "What is your purpose for entering our country? Are you Jewish?"

This again.

I smiled to myself. I thought about how his mannerisms, his tone, and even his questions sounded like my Iranian interrogator. I explained that I was traveling with my sweetheart, Jenny, and my father for ten days. We planned to visit my extended family. They are all Jewish, and all are Israeli residents and citizens.

He eventually found the Iranian stamp on the last page of my passport. When I explained why I had an Iranian stamp, he asked, "Do you have any contacts in Tehran?"

"No," I replied evenly.

The interrogator spent another few minutes on his computer. Then he made a couple phone calls in Hebrew. Eventually, he continued to question me. "Did the Iranian government ask you to gather intelligence for them?"

"That's funny," I said with a forced laugh. "When I was interrogated in Iran, they accused me of spying for you."

Thirty minutes later he let me go. I walked off with the slight discomfort that a sailor feels upon being sprayed with water from the sea. The wetness was annoying, but it was far from a shipwreck. My family had left the airport confused about why I didn't arrive on the plane I had boarded. Before Jenny and I hailed a taxi to my family's home in the suburbs of Tel Aviv, she spotted breaking news on TV: Israel had just started a war with Gaza.

Five days later, I was with my father and I was shouting, "Run! Come on, Dad! Run!"

The air raid sirens blared through the Old City of Jerusalem. We had about sixty seconds to take cover before expecting an explosion.

My father alternated between speed-walking and jogging, and he needed my encouragement. Jenny, my aunt, my father, and I were the only ones left in the open plaza one hundred yards from the Western Wall. I kept scanning the sky for a bomb to drop on us any second. With the sirens ringing, we tucked into a nearby stone tunnel where scores of other tourists and pilgrims were already holed up. In the relative safety of that tunnel, there was nothing to do but wait.

Until this trip to Jerusalem, I had bowed to my family's pressure to stay in their suburban enclave while the war on Gaza raged on. My cousins and uncles and aunts worried for my safety and wanted me to take extra precaution. One cousin thought I should just be grateful the Iranians didn't kill me and stay put. My mother was anxious. She had even feared I would get abducted by Iranian agents at the Newark airport before flying to Israel. And my father insisted that if I visited Jerusalem, he'd accompany Jenny and me — as if my dad's presence would stave off a (potentially Iranian-made) bomb from Gaza.

By my calculations, we'd be safe traveling within Israel, and I told them so. After all, of the approximate 180 deaths by rocket fire, over 95 percent of them were of Palestinians inside the Gaza Strip. For my family, it wasn't about statistics; it was about precaution. Moreover, I was in no position to convince them I'd be safe. It must have seemed to them like I hadn't learned anything. First I hiked in Iraq, and then I wanted to leave their suburb during the war. One family member told me that being an antiwar, pro-Palestinian leftist blinded me to life's dangers. Though I knew Kurdistan was safe and that the sirens in Jerusalem were a statistical improbability, I started to doubt myself.

Suddenly the sirens stopped. Relieved, we followed a Japanese tour group out of the archeological site to the Western Wall. I saw no trace of a bomb blast. At the wall, I inserted my prayers into one of its many cracks. I thanked God for keeping me safe, and then I lingered a bit. I stood there thinking about walls, barriers, and borders, and I wondered why an ancient wall is the central landmark of Judaism. I thought about fear and how hard I'd fought it in prison and how I still need to guard against a culture of fear.

. . .

Two days later, Egypt brokered a truce between Hamas and the Israeli government, which put an end to the war. The following day, Jenny and I drove to the Palestinian city of Bethlehem to meet up with a university professor who was a friend of a friend. To my family, I was willfully entering enemy territory. They thought I'd be treated as an enemy because of my heritage.

I rolled down the car window at an Israeli military checkpoint outside of Bethlehem. The soldier looked at me and asked, *"Yahudi?"*

This question reminded me again of my Iranian interrogations, just as it had at the airport. The lingering fear that I'd be treated differently than Shane and Sarah had scarred me — even though the interrogators, the guards, and the judge never treated me any differently on account of my religion.

My mind then flashed to Salem and the night Shane and I were released. On the flight from Tehran to Oman, he told me that an official in Iran suggested negotiating for Shane and me separately. Salem told me that he had immediately rejected the idea. "It would be nearly impossible to get the last one out," he explained. Salem then looked at me gently and said, "And we know who that would have been." I assumed it'd be me because of my heritage.

The nightmare of still being in Evin Prison gnawed at me. So, one day, many months after my release, I called Salem. He told me that Shane was the obvious candidate for release, but not because I was Jewish. Rather, Shane's reunion with Sarah would look beautiful in the media and make Iran seem compassionate.

Now, at the checkpoint, Jenny nudged me. I turned to the Israeli soldier waiting at my car's window. "Yes, I'm Jewish," I said, and he let us pass.

I drove ahead, partly relieved that I gave the soldier the answer he wanted. I was also embarrassed that I was awarded such privilege for my religion. Locked behind the wall that surrounds Palestinian territory, the professor offered me generous hospitality and showed me around his city. He patiently explained how his people struggle with the injustices of a forty-five-year military occupation. I understood the injustice in front of me, but — as I never had before my imprisonment — I felt it viscerally.

• • •

My trip with Jenny lasted ten days. We had been sharing our lives to-gether since I moved from Oakland to live with her in West Virginia, and I wanted her to meet my Israeli family.

I first fell in love with her in the seventh grade. I remember playing Ouija and hide-and-seek in the dark on weekends with her. During the school day, we'd talk about life. We decided it wasn't nice to hate people whom we didn't like, so we'd only "highly dislike" them. We thought our names worked better mixed together: I became Josh-ifer and she became Jenny-ua. For the first time, I fell in love. I eventually asked her to go out with me. Our "romance" lasted a few months and we rarely even held hands. We liked being friends better.

I recalled the details of our relationship in prison. We regularly jogged together before high school, we partnered together for after-school dance lessons, and I distinctly remembered burning a huge vat of couscous that we brought to social studies class for a presenta-tion on Middle Eastern culture.

We grew up. I went to the West Coast for college and she went to the South. One night in the summer when we were twenty-three and both happened to be in our hometown, I called Jenny to hang out. We went to a bar, shared a beer, and danced to salsa music. Eve-ryone else left the dance floor, but we danced until they closed down. Jenny agreed to come back to my place. We dipped pretzels in choc-olate and snuggled on the couch. "I've been wanting to kiss you for a long time," I told her. We laughed about how nervous we were in seventh grade. Then Jenny smiled at me and we kissed. That night, I confessed to her that I had told my best friend in high school that I couldn't imagine marrying anyone but her.

When I saw Jenny at my welcome home party after my release from prison, her joy on the dance floor drew me to her as it did years before. I watched her radiant eyes look at me throughout the night, and I knew that after all these years we were finally ready for each other.

After the party, we sat on a swing in my parents' backyard and talked for hours. I found out that I was in her thoughts just like she was in mine. She had been hunger-striking and sending letters and books that I never received. In those first months of freedom, I saw her whenever I could, and when I traveled around the country, we sent each other handwritten letters.

With Jenny, civilian life started making sense, and I felt more comfortable being myself. She encouraged me to follow my heart and my interest in history, law, yoga, and qigong.

A month after our trip to the Middle East, Jenny and I went for an ultrasound to confirm that she was pregnant. The doctor called me into the exam room. Jenny lay on her back, and the doctor showed me the monitor. The fetus's heart beat at a healthy 171 beats per minute. I looked into Jenny's beaming eyes and clutched her hand. We smiled.

Isaiah Azad Fattal was born the day before the four-year anniversary of my detention. As I watch him grow before my eyes, I'm heartened to witness life start anew. While wars, incarceration, and trauma cycle again and again, Isaiah Azad reminds me that there is always hope for a new world to be born: it is still possible that "nation shall not lift up sword against nation"; it is still possible to end the mutual hostility that led to our detention; and it is possible to find an old love that feels brand-new.

SHANE

The second night of freedom in Oman felt like the opposite of the first. I was in bed with Sarah and I couldn't sleep. My mind was moving fast, but erratically. Suddenly, it hit me: I was free, but I was not well. I broke down. Had I lost a part of myself, irrevocably? I felt mentally handicapped, as if my mind were enshrouded in a cloud. In prison, my mind was clear and sharp, but now it was an ice block melting in the sun. "It's okay, I'll take care of you," Sarah told me in the darkness, rubbing my chest. "It's just going to take a little time."

My brain couldn't handle all the input. I'd become accustomed to dealing with only one stimulus at a time — talking with Josh, the sound of footsteps, the book I was reading. Now I felt like an infant — simple, vulnerable, and lost to the complexities of the world. When taken to a restaurant and given a menu, I couldn't choose what to eat. I was also unable to read many of the subtleties of anyone's body language except Josh's, and to a lesser extent Sarah's. I began in-structing people to tell me what they meant in precise words. Free-dom was a foreign country to me.

A couple days after we gave a press conference in New York, our lawyer, Masoud Shafii, set out to visit his daughter and son-in-law in the United States. As he was about to board the plane in Tehran, the authorities pulled him aside, took his passport, and told him he couldn't have it back. He wasn't allowed to travel. A wall was erected between him and his family. He left the airport and went home. The next day, he was arrested.

When I first heard about this, I pressed my head into my knees and wept with a force I hadn't experienced since our early days in prison. I felt as though my freedom had been revoked, as though Josh and I had merely traded positions with someone else. It felt like the Iranian government was punishing us with someone else's captivity because we hadn't praised them in the media.

Shafii was released after a day of interrogation. His life has not been the same since. The court refuses to take any of his cases. Intel-ligence agents told him they had a case open on him, a thinly veiled threat to punish him should he step out of line.

As his life became rapidly worse, TV news shows played portions of our New York press conference along with footage of Josh and me coming off the plane and into the arms of our families. The cover-age was selective. The major networks aired our most graphic de-scriptions of the conditions of Evin Prison, but they left out the parts where we said that when we complained about those conditions, guards reminded us of comparable conditions in Guantánamo Bay and CIA black sites around the world. The human rights violations on the part of our government in no way justified what was done to us, we said, but they provided governments like Iran an excuse to act in kind. No major network aired any of that.

President Obama stated publicly that he was "thrilled" with our release, but one former government official was less excited. Elliott Abrams, a high-level diplomat of both the Reagan and George W. Bush administrations, condemned what he called my "ingratitude." He was referring to the statement I made after landing in Oman, call-ing for the release of "other political prisoners and other unjustly imprisoned people in America and Iran." Mr. Abrams said the statement left a "bad taste" in light of the "immense diplomatic activity this country undertook" to free us. He found insulting the sympathy of someone who was wrongfully detained toward others in his position.

He also believed the premise of the statement to be false. In his post on CNN.com the day after our release, Mr. Abrams asked, "Who exactly are the 'unjustly imprisoned people in America'? . . . Can we have some names?" I suspect many others had the same question. There was a sad irony in the timing of Abrams's question. On the very day Josh and I were released, a man named Troy Davis was executed in Georgia. His case rested on the testimony of seven witnesses, five of whom later recanted. Some said they'd been coerced by police to implicate Davis. President Jimmy Carter, Amnesty International, former FBI director and judge William Sessions, and many others called on the courts to grant Davis a new trial or evidentiary hearing. They refused, and executed him instead. Situations like his are not uncommon. The most conservative studies say that ten thousand people are wrongfully convicted of serious crimes every year in the United States. The majority of these are probably African American; the tiny fraction who are exonerated, sometimes decades after their conviction, are overwhelmingly black.

Though we don't generally admit it, we do have political prisoners in the United States. One example is Bradley Manning. Manning was imprisoned for leaking hundreds of thousands of diplomatic cables as well as army reports related to the wars in Iraq and Afghanistan. It is true that in doing this, he broke the law.* When he first came across evidence that the Iraqi government was imprisoning people for criticizing the prime minister, his superior officer reportedly ordered him to keep it quiet. He quickly learned that, when it came to reporting military abuses, official channels were fruitless. He broke the law because it was the only way for him to do what he believed was right.

---

* Manning has since changed his gender to female and is now known as Chelsea. I use the pronoun "he" here since it is how she referred to herself at the time these events took place.

Manning's case is nothing if not political. He has been sentenced to thirty-five years in prison for *leaking* information that was being kept from the public. Many of those implicated by information he leaked do not face legal action. To give one small example, Manning's leaks revealed military documents that implicated the private American military contractor Dyncorp in the crime of child trafficking. The documents show that they paid for an underage male sexual slave to dance for some of their Afghan employees. None of them have been brought to court. I struggle to find anything other than a political explanation for the disparity in the handling of these two instances of law breaking.

Perhaps the most salient example of unjust detainment in this country is not on U.S. soil, but at Guantánamo Bay. Prisoners at Guantánamo have complained of beatings, sleep deprivation, prolonged hooding, and sexual humiliation. It has been well established that some were tortured, sometimes extensively, before being transferred to Guantánamo. As I write this, 164 people are being held there without trial. Forty-six have been officially designated for indefinite detention without charge or trial by the Obama administration. Eighty-four others have been approved for release but remain, years later, in detention.

In Evin, we frequently challenged guards, prodding them to come to terms with the blatant injustice of the system they supported. Not a single guard ever accused us of being spies, but when pushed, most would justify our imprisonment by saying that we crossed the border illegally, that we broke the law. "Did we?" we'd ask them. "How do you know when we've never been tried in court?"

The right to be brought before an impartial court in a timely manner is the very basis of a fair criminal justice system. That right does not exist in Iran. It does not exist for prisoners held in Guantánamo Bay either. That right has not existed for some immigrants in this country since at least the passage of the Patriot Act in 2001. Three months after our release, President Obama signed into law the right of the United States to hold people without trial, including U.S. citizens, who are suspected of the extremely broad classification of "involvement in terrorism." In doing so, he essentially approved for the

United States what he and Secretary Clinton implicitly condemned in Iran's handling of our case: the authority of a government to detain people without due process. He even took it a step further. We know (because of government leaks) that our president keeps a list of people whom he has authorized our security services to kill without trial, including U.S. citizens.

We are fortunate that we had so much support. Most people coming out of prison do not. When I look back over the past year and a half since our release, I see the glances of recognition from strangers on the streets and innumerable instances of people I didn't know welcoming me home. I see months full of firsts — my first run outside, first home-cooked meal, first time hiking, first ice cream cone, first time getting rained on. I see the slow process of relief my family experienced. I feel the deep comfort of my mom's fireplace. I see my wedding with Sarah, nestled amidst redwood trees on the California coast. I see our beautiful apartment and our cat resting in one of its many patches of sun.

I also see a long period when I carried a tension inside, one that pushed out against my skin and strangled my lungs at the same time. Some nights, I dreamed of returning to prison. Other nights, I dreamed of escaping it. I dreamed of Dumb Guy over and over again. I screamed at him. I punched him. I spewed out something deep and pent-up at him. Outside of those nightmares, in my daily routine, I became unable to sit still. I often felt trapped. I got nervous in crowded buildings. I got depressed when I was alone. Sometimes, at the end of the day, I felt like it was impossible to get enough air. Sometimes I wished I had a prison guard to fight. Sometimes I yelled at Sarah as though she were a guard. I was out of prison — I knew that — but I couldn't convince my whole body it was true. Fortunately, that period also slowly faded away.

When I look back, I see one moment in which my awareness was probably the clearest it has ever been in all my life: coming off that plane in Oman. I knew precisely what was most important to me then: the people I love, my freedom, and the freedom of others. There were no second thoughts or judgments. There was no fear of retribution.

It is strange to have cameras present at such a moment, to have the world bear witness to such a profound personal experience, but I am glad they were there. I am glad that people could share the power of a moment of freedom. And I am grateful that in the midst of all this, I was able to say the one thing that made the most sense to say: that everyone deserves justice. Everyone.

ACKNOWLEDGMENTS

Dozens of people dedicated the better part of two years of their lives to getting us out of prison, and tens of thousands stood behind them. It's impossible here to name even a fraction of them, but our gratitude to each of them is deep. We thank everyone who offered their support, in all of the forms it took.

First, the three of us are lucky to have three devoted, selfless, and resourceful families. Every one of our family members poured their lives into obtaining our freedom, as well as helping us heal and adjust when we came back home. We can never repay what they have given to us. Our deepest gratitude goes to Josh's parents Laura and Jacob Fattal, his brother Alex Fattal, his grandparents Muriel and Carroll Felleman, and his uncle Fred Felleman; Sarah's mother Nora Shourd, her sister Martha Webster, her brother Chris Rapp, her father Louis Shourd, her aunts Karen and Suzy Sandys and uncle Mike Sandys; Shane's mother Cindy Hickey and father Al Bauer, his stepfather Jim Hickey, his sisters Nicole Lindstrom and Shannon Bauer, their spouses Nate and Natalie, his grandparents Mary and Alfred Schmidthuber and Lila Bauer, and his aunt Cathy Theis. We would also like to thank all the rest of our aunts and uncles, our nieces, nephews, and cousins. Every one of you made a difference, each in your own unique way.

Second, we'd like to thank our incredible lawyer Masoud Shafii and his loving, supportive family for their unparalleled sacrifice. Special thanks to Sultan Qaboos bin Said of Oman, his devoted envoy Salem al-Ismaily, and Ambassador Hunaina al-Mughairy — without your brilliant diplomacy we might still be in prison today. Others who played crucial roles in negotiating our release include the Swiss ambassador Livia Leu Agosti and her staff, the Iraqi president Jalal Talabani and his son Qubad, Sean Penn, the late president Hugo Chávez of Venezuela and the current president Nicolas Maduro, Fernando

Sulichan, and Temir Porras Ponceleon. Brazil, Turkey, Greece, and
other nations weighed in for our case to be resolved. It is a rare, beautiful thing when powerful people use their positions and resources so
selflessly, and for that we owe you our admiration.

The Free the Hikers campaign was held together by a close-knit
pool of individuals who worked wonders on our behalf. We would
like to give special thanks to our dear friend Shon Meckfessel, who
fortunately did not hike with us on that fateful day. You sprung into
action immediately, letting the world know what happened to us,
and remained devoted until the end of the ordeal. Along with Shon,
we'd like to thank the campaign's central organizers: Jennifer Miller,
Liam O'Donoghue, Tegra Fisk, Laura Brechheimer, and Ben Rosen-
feld. Thanks also to all the tireless fundraisers and donors, artists and
performers, photographers and graphic designers, letter writers and
book senders, protesters and organizers, in particular Joey Boxman,
David Martinez, Moriah Oxnard, Leigh Goldenberg, Bessa Kautz,
Margaret Roberts, Brian Gralnick, Darryl Wong, David Brazil, Sara
Larson, David Keenan, Sami Feld, Deanna Tibbs, Rebecca Fischer,
Jennie Hienlien, Lia Rose, Emily Churchill, Rachel Clearwater, Pau-
line Bartolone, Scott Rabinowitz, Emily Sudd, Meredith Walters,
Ethan Rafal, Sarah Hobstetter, Cait Quinlivan, James Tracy, Marcus
Kryshka, Will Kabat-Zinn, Michelle Borok, Andrea De Moral, Dave
Petrelli, Qilo Matzen, Kristina Lim, David Marcus, Salina Abji, Moxie
Marlispike, Ryan Harvey, Vassia Alaykova, Bob Stein, May Abdalla,
Kristof Cantor, Dina Solomon, Laura Allen, Peter Ralph, Ian Mac-
Kenzie, Jafar and Chris Saidi, Matthew McNaught, Elizabeth Sy, Eric
and Riba Lendle, Marilena LoVerde, June McIntyre, Michele Bloom-
berg, Sage Warren, Patty Wrightson, Mia Nakano, Melanie Pickrell,
Shane Bauer of Duluth, Minnesota, Hilary Klein, Heyward, James
Sadri, Scott Campbell, Tristan Anderson and his parents, Gabrielle
Silverman, Cindy and Craig Corrie, David Rhode, Bonnie Abaunza,
Anna Baltzer, Sarah and Josh's former students, and many, many
more. Your tremendous support sustained the campaign for our
freedom. Many of you gave critical emotional support to our families
during our detention and to each of us during our often difficult re-
adjustments to free life.

Our unending gratitude goes to our communications specialist Paul Holmes and his family. Also to our brilliant, devoted social media and website experts Farah Mawani and Alita Holly. Thanks to our translators: Pari (pseudonym), Shirin (pseudonym), Pouria Montazeri, and Sadeq Rahimi. Thanks to our advisors Gary Sick, Ambassador Pierre Prosper, Karim Sadjampour, Stephen Zunes, Ambassador Bill Miller, Trita Parsi, Dr. Akbar Ahmed, Aleen Stein, and others. Thanks to the publicist of the Free the Hikers campaign, Samantha Topping, and our videographers and filmmakers Jeff Kaufman, Bobby Field, and Natalie Avital. Thanks to Roxana Saberi, Haleh and Shaul Esfandiari, Terry Waite, and Eric Volz for extending their experience, understanding, and compassion about how to deal with wrongful detention. We'd like to thank the U.S. State Department's Jake Sullivan, Undersecretary of State Bill Burns, and the late Philo Dibble for their special attention. Our appreciation also goes to consular services for faithfully passing on our books and letters. Thanks to Ambassador Richard Schmierer and his wife Sandy for their work on our behalf and their wonderful hospitality during our two trips to Oman.

We'd like to thank all the religious leaders around the world who advocated for our release, in particular Cardinal Theodore Edgar McCarrick, Imam Hassan al-Qazwini, Reverend Nora Smith, Imam Magid, Bishop Chane, Reverend Anne Hansen, and the late archbishop Pietro Sambi. We appreciate those who relentlessly pushed our cause in Congress, including several senators and representatives and their staffers. Many thanks to countless embassy officials and organizations such as Amnesty International, Safe World for Women, Human Rights Watch, Community to Protect Journalists, International Honors Program, Friends of the Earth, Search for a Common Ground, Innocence Project, National Iranian-American Council, Islamic Society of North America, Council on American-Islamic Relations, Care2, Intent.com, Witness, and the United Nations. We'd also like to thank all the bloggers and journalists who accurately and faithfully covered our story, in particular Amy Goodman and her colleagues at Democracy Now! Thanks to Shane's editors Monika Bauerlein and Clara Jeffery at *Mother Jones*, Sandy Close at *New*

America Media, Richard Kim at the *Nation*, and Esther Kaplan at the Nation Institute for standing by us and supporting our families. We deeply appreciate the public advocacy from Muhammad and Lonnie Ali, Noam Chomsky, Desmond Tutu, Yusuf Islam, Tom Morello, Mairede McGuire, Deepak Chopra and his family, President José Ramos-Horta, and Ban Ki-moon.

Last, we'd like to give a shout-out to our amazing friends in numerous cities around the world: Athens, Atlanta, Baltimore, Bangalore, Berkeley, Boston, Boulder, Burlington, Cape Town, Changsha, Cottage Grove, Damascus, Duluth, Elk Grove, Houston, Hultsfred, Irvington, Johannesburg, Laguna Beach, London, Los Angeles, Lunenburg, Medellín, Minneapolis, New York, Norwich, Oakland, Ottawa, Paris, Philadelphia, Pine City, St. Paul, San Francisco, Sarasota, Seattle, Shakopee, Shanghai, Tehran, Toronto, Vancouver, Washington, DC, Williamstown, Worcester, and many more.

Our book could not have been written without the dedication of our agent Bill Clegg of WME Entertainment, the support of our lawyer Eric Rayman, and the clear vision of our brilliant, patient editor Eamon Dolan of Houghton Mifflin Harcourt. We also want to thank a handful of friends and colleagues who helped us with the fact-checking and translations needed to complete the manuscript: Roya Boroumand, Amir Soltani, Firuzeh Mahmoudi, Sadeq Rahimi, Trita Parsi, Sima Alizadeh, and Abbas Hakimzadeh.

Special thanks to Rafaella Cohn, Jenny Bohrman, Mary Helen Spetch, Ben Christopher, and Maraya Karena for looking over Josh's scenes and providing invaluable feedback. Josh wants to especially thank Jenny for tirelessly supporting him through seemingly endless deadlines and his rollercoaster emotions. Fred Felleman, Beth Miller, Cici McLay, George McKinley, Maria Kelly, Devon Bonady, and the Aprovecho family provided generous hospitality and beautiful settings in which to write.

Though many lives were shaken and forever changed by the events recounted in these pages, the purview of this memoir is limited to the experiences of its three authors. The story of the Free the Hikers campaign, especially in the first year, could be a separate book. Three names in this book — the prisoner "Hamid," the guard "Ehsan," and the translator "Pari" — have been changed to protect the safety of those individuals. The names used for other guards and interrogators are either the names they used for themselves or names we assigned to them in prison because they kept their names secret.

Every time a person is imprisoned, the ripple effects are enormous. When people join together to fight injustice, the results are equally enormous. Thank you all for giving us our lives back.

ABOUT THE AUTHORS

JOSHUA FATTAL is a historian. Prior to his arrest in Iran, he taught in Asia about the political economy of healthcare and was codirector of an environmental education center in Oregon. Joshua has also taught nonviolent communication, qi gong, and yoga. He lives in Brooklyn, New York, with his partner and child.

SARAH SHOURD is a writer, educator, and contributing editor at *Solitary Watch* currently based in Oakland, California. After her wrongful imprisonment in Iran, she has focused her human rights advocacy work on combating the widespread use of prolonged solitary confinement in U.S. prisons and jails.

SHANE BAUER is an award-winning journalist and senior reporter at *Mother Jones* magazine. His work has appeared in the *Nation, Salon,* the *Los Angeles Times,* the *Guardian,* and many other publications. He lives in Oakland.